NATO ASI Series

Advanced Science Institutes Series

A series presenting the results of activities sponsored by the NATO Science Committee, which aims at the dissemination of advanced scientific and technological knowledge, with a view to strengthening links between scientific communities.

The Series is published by an international board of publishers in conjunction with the NATO Scientific Affairs Division

A Life Sciences **B Physics**	Plenum Publishing Corporation London and New York
C Mathematical and Physical Sciences **D Behavioural and Social Sciences** **E Applied Sciences**	Kluwer Academic Publishers Dordrecht, Boston and London
F Computer and Systems Sciences **G Ecological Sciences** **H Cell Biology**	Springer-Verlag Berlin Heidelberg New York London Paris Tokyo

The ASI Series Books Published as a Result of Activities of the Special Programme on CELL TO CELL SIGNALS IN PLANTS AND ANIMALS

This book contains the proceedings of a NATO Advanced Research Workshop held within the activities of the NATO Special Programme on Cell to Cell Signals in Plants and Animals, running from 1984 to 1989 under the auspices of the NATO Science Committee.

The books published as a result of the activities of the Special Programme are:

Vol. 1: **Biology and Molecular Biology of Plant-Pathogen Interactions.** Edited by J. A. Bailey. 1986.

Vol. 2: **Glial-Neuronal Communication in Development and Regeneration.** Edited by H. H. Althaus and W. Seifert. 1987.

Vol. 3: **Nicotinic Acetylcholine Receptor: Structure and Function.** Edited by A. Maelicke. 1986.

Vol. 4: **Recognition in Microbe-Plant Symbiotic and Pathogenic Interactions.** Edited by B. Lugtenberg. 1986.

Vol. 5: **Mesenchymal-Epithelial Interactions in Neural Development.** Edited by J. R. Wolff, J. Sievers, and M. Berry. 1987.

Vol. 6: **Molecular Mechanisms of Desensitization to Signal Molecules.** Edited by T. M. Konijn, P. J. M. Van Haastert, H. Van der Starre, H. Van der Wel, and M. D. Houslay. 1987.

Vol. 7: **Gangliosides and Modulation of Neuronal Functions.** Edited by H. Rahmann. 1987.

Vol. 9: **Modification of Cell to Cell Signals During Normal and Pathological Aging.** Edited by S. Govoni and F. Battaini. 1987.

Vol. 10: **Plant Hormone Receptors.** Edited by D. Klämbt. 1987.

Vol. 11: **Host-Parasite Cellular and Molecular Interactions in Protozoal Infections.** Edited by K.-P. Chang and D. Snary. 1987.

Vol. 12: **The Cell Surface in Signal Transduction.** Edited by E. Wagner, H. Greppin, and B. Millet. 1987.

Vol. 19: **Modulation of Synaptic Transmission and Plasticity in Nervous Systems.** Edited by G. Hertting and H.-C. Spatz. 1988.

Vol. 20: **Amino Acid Availability and Brain Function in Health and Disease.** Edited by G. Huether. 1988.

Vol. 21: **Cellular and Molecular Basis of Synaptic Transmission.** Edited by H. Zimmermann. 1988.

Vol. 23: **The Semiotics of Cellular Communication in the Immune System.** Edited by E. E. Sercarz, F. Celada, N. A. Mitchison, and T. Tada. 1988.

Series H: Cell Biology Vol. 23

The Semiotics of Cellular Communication in the Immune System

Edited by

Eli E. Sercarz

Department of Microbiology, University of California
405 Hilgard Avenue, Los Angeles, CA 90024, USA

Franco Celada

University of Genova
Transplantation Immunology Service, Hospital S. Martino
Viale Benedetto XV, 20, 16132 Genova, Italy

N. Avrion Mitchison

Tumour Immunology Unit, University College London
Gower Street, London WCIE 6BT, England

Tomio Tada

Department of Immunology, Faculty of Medicine, University of Tokyo
Tokyo, Japan

With Doodles by Umberto Eco

Springer-Verlag
Berlin Heidelberg New York London Paris Tokyo
Published in cooperation with NATO Scientific Affairs Division

Proceedings of the NATO Advanced Research Workshop on The Semiotics of Cellular Communication in the Immune System held at Il Ciocco, Lucca, Italy, September 9–12, 1986

ISBN-13:978-3-642-73147-1 e-ISBN-13:978-3-642-73145-7
DOI: 10.1007/978-3-642-73145-7

Library of Congress Cataloging-in-Publication Data. NATO Advanced Research Workshop on the Semiotics of Cellular Communication in the Immune System (1986 : Lucca, Italy) The semiotics of cellular communication in the immune system / edited by Eli E. Sercarz ... [et al.]. p. cm.—(NATO ASI series. Series H, Cell biology : vol. 23) "Proceedings of the NATO Advanced Research Workshop on the Semiotics of Cellular Communication in the Immune System held at Il Ciocco, Lucca, Italy, September 9–12, 1986"—T.p. verso. "Published in cooperation with NATO Scientific Affairs Division.
ISBN-13:978-3-642-73147-1 (U.S.)
1. Immune system—Physiology—Congresses. 2. Immunocompetent cells—Congresses. 3. Cell interaction—Congresses. I. Sercarz, Eli E. II. North Atlantic Treaty Organization. Scientific Affairs Division. III. Title. IV. Series. [DNLM: 1. Cell Communication—congresses. 2. Immune System- -physiology—congresses. QW 504 N279s 1986] QR182.N38 1986 616.079—dc 19 DNLM/DLC for Library of Congresses
88-20189

2131/3140-543210 – Printed on acid-free paper

Editors' Preface

This volume contains the contributions to the workshop "The Semiotics of Cellular Communication in The Immune System" which took place at "Il Ciocco" in the hills north of Lucca, Italy, September 9-12, 1986. The workshop was the first meeting of what we hope will be a broad consideration of communication among lymphocytes, and focused on the new interdisciplinary branch of biological sciences, immunosemiotics. It is in the realm of the possible, if not the probable, that in the future a number of scientists larger than the thirty present at Il Ciocco will find immunosemiotics to fill a need in scientific thinking and a gap between biology and the humanities. This might lead to growth and flourishing of the branch, and in this case the first conference and this first book could be blessed by the impalpable quality of becoming "historical", if in an admittedly limited sense.

Just in case this should happen the organizers/editors think it wise to set the record straight at this particular time, about the sequençe of events and circumstances that crystallized the archeology of the "Il Ciocco" gathering. They feel a sort of obligation to this endeavor: it has happened all too often that innocent historians have been left in utter confusion by the careless founders of new religions, schisms, revolutions, et cetera, who simply forget to jot down the facts before the whirlwind of time engulfs them in its fog.

One noon in March of 1984, Celada, Mitchison and Sercarz left Los Angeles aboard a Volkswagen minibus, heading towards Park City, Utah. They were going to attend a UCLA conference on Regulation of the Immune Response. Sercarz - quite typically - was late in preparing his manuscript for the meeting, which had to be about T-B interactions and the signals that made it possible, and wanted to use the time of the trip for the preparation. He was haunted by the term immunosemiotics since Tomio Tada had put it in print a couple of years before as an approach that could unravel the intricacies of communication between lymphocytes. Sercarz needed to know more about semiotics in general: not having located a notable book in the library, he had urged Celada to take with him for the trip the "Trattato di Semiotica Generale" by Umberto Eco[2] - an Italian edition found on the bookshelf of Celada's son Luca, who happened to be a student of communication at UCLA.

The odds were against the possibility of a serious scientific discussion in the loud van, but the odds were defeated by the unexpected, as always: this time, the finding of a string of "no vacancy" signs on all motels lining the road to Utah. This obliged the three to undergo an epic 17 hour drive without interruption, and gave tnem ample time. Sercarz questioning, Celada frantically reading the relevant passages of the text and translating them into a broken English. Mitchison, scholarly reformulating the issues in the real language and acting as a happy father of cell cooperation; he was also particularly enthusiastic about Eco, having just finished the Name of the Rose, this a bestseller in England. Facts were listed, speculations launched about similarities of approach but also the compatibility between the experimental method and the entities of semiotics. If antigen is the signifier, where is the signified? But if the T helper cell is the signifier, is antigen signified? And who is to interpret the signs we, the immunologists or the immune system, through its components and cells? On this last point, a heated discussion flourished, which was to be carried over to the workshop (it is clear now with later knowledge - il senno di poi - that immunosemiotics prefers the second alternative, the talk among the components of the immune system. While immunologists and semiologists alike look upon the intercellular communication they do not participate.) At intervals it was necessary to refuel. A square wheel of ripe Taleggio cheese that one of the three (F.C.) had just smuggled from Milano provided nourishment during the many stops at the service stations of the desert and contributed to the cozy atmosphere inside the VW. The discussion continued. The three men became bolder, while the hours and the miles went by. By the time they arrived at the Great Salt Lake it was apparent to them that immune cell interaction was a perfect arena for application of the broad methods of semiotic analysis to a current problem in biological recognition. The cells of the immune system provide an unparalleled opportunity for the study of biological meaning, and of the utilization of signs and signals in regulating intercellular commerce.

The stage was surely set for such an exploration. In the last ten years, there had been a explosion of information about the structure of molecules and factors secreted by immune cells from the two major arms of the immune system, the T and B cells. First, we know a large amount of detailed information about the molecules serving as receptors for the external world of antigens. Second, because of the unique aspect of the immune system that receptors on T and B cells are also directed toward the recognition of each other (the internal universe) as well as to the external universe, an intercellular web of recognition potentially joins all lymphocytes. This is the area of idiotypic interactions. Third, the number of known cell subpopulations with different functions has increased greatly. This has led

TABLE OF CONTENTS

to one view that a single cell can only perform a single function, but also has stimulated the counter-argument that apparently contradictory functions might be carried out by the same cell during its lifespan.

Finally, the repertoire of recognition of different functional lymphocyte subsets presents interesting problems in recognition: basically, how can discrete T and B cellular families, each collectively capable of recognizing on the order of 10^7 different structures, but having to collaborate with each other to induce an immune response, ever specifically find appropriate partners within the organism? What the the rules allowing unambiguous interactions between these partners? The adaptability of the system is such that there is an enormous compositional diversity, also allowing for a rich semantic content.

In summary, the interactions of a particular lymphocyte in one subpopulation with the appropriate partner from another subpopulation for regulatory or cooperative purposes requires an efficient and unambiguous system of intercellular recognition and pairing. This problem can be seen as an immunosemiotic one, involved the perception of specific signs by each of the cellular participants. Tomio Tada framed the definition of this science as the study of the general principles underlying the structure of sign systems perceived by different cells of the immune machinery. Thus, restrictions in partner cell interactions must exist as part of an intercellular semiotic system.

When the minibus started to engage the winding curves of the Utah mountains, and snow began to appear on the ridges of the Rocky Mountains, the idea of the Il Ciocco workshop emerged naturally. The objective was to be an attempt to establish rules for the organization of cellular interactions by thorough analysis of the signs which are displayed and recognized by different members of the cellular community. The ultimate hope was to understand the ways by which various restriction elements are utilized in the system, the ways in which a maturing lymphocyte might receive and send messages, and the safeguards in the system to prevent undue complexity. There even was hope that as the underlying rules of cellular communication were learned, an elegant simplicity would be revealed.

Whether these purposes and dreams were realized, even in a small proportion in the workshop is very difficult to decide. The meeting surged with high expectations, punctuated by declarations of various immunologists as well as the irrepressible Umberto Eco, all hoping to establish the framework of future discussion. We were treated to the delightful violin/piano musical translations of the genetic code by Susumu Ohno, we witnessed on video the opera staged by Polly Matzinger _et al_. at the Basel Institute after their disastrous fire; and enjoyed the playful meanderings of the noted canine immunologist, Gabraliel Mirkwood. It was a heartening sign that the participants wanted the organizers to set a time for a second workshop. But the acid test will be the reading of this volume, where much to the delight of the organizers, it appears that the seeds planted during this novel get-together bore many thoughtful fruits.

The organizers would like to thank the entire NATO group, that provided support for the meeting, Rabyn Blake who chronicled the proceedings photographically with characteristic verve, and the extremely helpful and cordial staff at Il Ciocco. We look forward to the next episode in this saga.

À L'OMBRE DES LYNPHOCITES EN FLEUR

ON SEMIOTICS AND IMMUNOLOGY

Umberto Eco
Universita degli Studi di Bologna
Institute di Discipline della Communicazione
Bologna, Italy

At the beginning of our meeting in September 1986 I felt the duty of stating many reasons that, according to my personal theoretical standing and to my very elementary notions about immunology, discouraged a direct application of semiotics to immunology.

After the meeting I have changed my mind. I feel still unable to say whether semiotics can help immunology, but I discovered that immunology can help semiotics.

If such a result can appear unsatisfactory for my immunologist friends, it seems very fruitful to me. By a happy case of serendipity, our meeting did something for the Advancement of Learning.

In the first part of this paper I shall summarize my introductory arguments. Since in the meanwhile Patrizia Violi has presented her paper, I shall take for granted many things she said about the requirements to be met in order to define any system as a semiotic one. I shall simply develop some other points of view on the same subject, complementary to her own.

In the second part of my paper I shall develop some remarks inspired to me by the discussions with my fellow immunologists.

I was very prudent in the course of my introductory presentation at the symposium because for a long time semiotics has been considered as an imperialistic discipline aiming at capturing every aspect of the world. I think that many aspects of the world, event though they are not semiosic in themselves, can be approached from a semiotic point of view. But it is equally important to ascertain what semiotics can capture and what - as far as I understand - it cannot do.

My opinions do not represent the opinion of the semiotic community at large: some of my colleagues are more optimistic and more catholic in their approach. However in my 'Theory of Semiotics' I have traced an upper and lower threshold of semiotics, maintaining that, as far as I was concerned, semiotics had to deal only with the subject matter lying in between.

1.1. Semiosis and semiotics.

First of all I want to make a distinction between semiosis and semiotics. Semiosis is a phenomenon; semiotics is a theoretical discourse

NATO ASI Series, Vol. H23
The Semiotics of Cellular Communication in the Immune System
Edited by E.E. Sercarz et al.

on semiosic phenomena.

According to Ch. S. Peirce (Collected Papers, 5.484), semiosis is "an action, an influence, which is, or involves, a cooperation of three subjects, such as a sign, its object and its interpretant, this tri-relative influence not being in any way resolved into actions between pairs." Semiotics is on the contrary "the doctrine of the essential nature and fundamental varieties of possible semiosis" (C.P., 5.488).

What is important in Peirce's definition of semiosis is that it does not take into account any interpreter or conscious subject. To summarize roughly the whole story, one witnesses a semiosic process when:
(i) a given an object or state of the world (in Peirce's terms, the Dynamical Object) can be translated into an interpretant, that is, another representamen.
(ii) is represented by a representamen and
(iii) the meaning of this representamen (in Peirce's terms, the Immediate Object) can be translated into an interpretant, that is, another representamen.

The Dynamical Object can even be an ideal or imaginary object or a state of a merely possible world. When represented it can be and usually is out of the range of our perception.

The representamen is a material expression such as a word or any other sign - better, it is the general type of many producible tokens of that sign.

The interpretant can be a paraphrasis, an inference, an equivalent sign belonging to a different sign-system, a whole discourse and so on.

In other words, there is a semiosic phenomenon when, within a given cultural context, a given object can be represented by the term rose and the term rose can be interpreted by "red flower", or by the image of a rose, or by a whole story telling how to grow roses.

1.2. Signification and communication

The interpreter - as the active protagonist of the interpretation - is certainly presupposed in the course of a process of communication (I tell <u>rose</u> to somebody and this somebody 'understands' that I mean "red flower"). But this interpreter is unnecessary in a system of signification, that is, in a system of instructions that makes "red" flower correspond, as a correct interpretant, to the expression rose.

I read in immunological papers many terms that certainly are concerned with semiotics, namely, biological meaning, recognition, communication, sign system, syntax, grammar, sign perception, and so on.

In light of my previous remarks it should be clear that some of these expresions refer to a communication process and some other to the existence of a sign-system. These two problems must be carefully distinguished.

There can be a sign-system that nobody actually uses in order to communicate (a private language designed for experimental purposes, or a new kind of Esperanto that nobody accepts) and there can be a communication process that takes place without a prexisting sign system: such would be the case of a European explorer A who starts interacting with a native informant B, where A guesses what B has in his mind by a trial-and-error process, but being however unable to design the 'code' that made their interaction possible.

1.3 Systems and semiotic systems

I found the term 'system' frequently used. but the notion of system is wider than the one of sign-system.

Let me consider the case of 'syntactic systems'. In its simplest form a syntax is an algorithm that generates strings of elements and discriminates between the acceptable and the unacceptable ones.

Let me imagine a syntactic system ALPHA considering a set of elements, a combination rule and three restrictions.

Elements: +, -, *, %
Rule: a string is composed of no more and no less than 3 elements.
Restriction 1. + can never be preceded by a -.
Restriction 2. neither * nor % can be followed by a sequence of the type +- or -+
Restriction 3: no element can appear twice in the same string.

The following are examples of acceptable strings:

+ - %
% * +
+ - *
+ % -

The following are examples of unacceptable strings:

* + -
% - +
+ + -
- + *

Such a syntax can rule different phenomena, for instance the growth of a tree. We say that the growth of a tree obeys syntactical rules. But we cannot say that the growth of a tree obeys the rules of a sign-system, because a sign-system is composed of a syntactics and a semantics as well (not to speak of possible pragmatic rules).

In order to have a sign-system one must match strings of the syntactic system ALPHA to strings of another system (for instance, a system BETA comprised four 'elements'):

+ - % stands for Water
% * + " " Fire
+ - * " " Air
+ % - " " Earth

(see in Violi's paper more technicalities on semiotic systems: in any case let's consider the ALPHA system as the Expression of which the BETA system is the Content).

1.4 Interpretation.

When one matches a syntactic system with a semantic one every string permitted by the syntactic system can be interpreted. It can be interpreted because one can say that + - % means "water", but also because "water" can be interpreted in its turn as "H_2O", as "a transparent drinkable liquid", by a specimen of water, or by an image representing water. In a semiotic system any content can become at is own turn a new expression that can be interpreted, or substituted, by another expression.

I want to stress that in interpretation, beyond the fact that (i) an expression can be substituted with its interpretation, it also happens that (ii) this process is theoretically infinite or at least indefinite and that (ii) when using a given sign-system we can both refuse to interpret its expression or to choose the most suitable interpretations according to different contexts.

In my previous writings - such as A Theory of Semiotics - I said that a sign-system is governed by reversibility: water stands for H_2O as well as H_2O stands for water. I maintained that such reversibility was distinguishing semiosic phenomena from mere stimulus-response phenomena. In my last writing (for instance Semiotics and the Philosophy of Language) I see reversibility only as a sort of 'optical' effect due to the fact that in semiosis every content can become at its turn the expression of a further content and that both expression and content can turn upside down by exchanging their roles. But by knowing that water means H_2O I learn something different from knowing that H_2O means water. In each case the

Immediate Object is interpreted under two different 'respects' or descriptions, or points of view.

I said that if a syntactic system ALPHA rules the growth of a tree this does not mean that ALPHA is a sign system. One could object that if we know the rule governing the growth of a tree we can infer the age of a tree from one of its sections. In fact, it is a semiotic principle (or at least a principle of my semiotics) that every phenomenon can be seen as a semiotic one as far as we take it as the sign of something else (for instance we can infer 'if smoke then fire' - where smoke is taken as the sign of an otherwise imperceptible fire). But to say that every phenomenon can be taken as a semiotic one does not mean that every phenomenon is a semiotic one. I can certainly say that if a dog wags its tail this means that it is happy, or that if I see red spots on the face of a person this means the person has measles: but neither the dog nor the person follow the rules of a sign-system. If there is a sign-system, it belongs to my competence and represents a semiotic rule I use to interpret events as if they were communicating something to me. I guess that if an immunologist sees (if possible) a given lymphocyte doing so and so, he will be able to predict that something so and so will happen or had happened. But such a principle is common to all scientific research as well as to normal experinence in our everyday life. If I see a rock falling on the head of my friend John, I can predict that John will feel bad: in interpreting a given sequence of events as the sign of their possible effects I behave semiotically; however neither John nor the rock are doing so. What Constantin Bona says in his paper must be stressed: one must distinguish between a semiotics of the immunological dictionary and the semiotics of the immune system.

In this sense the semiotics of the genetic dictionary knows that, in the passage between DNA and messenger-RNA

A -> U

T -> A

G -> C

C -> G

but it is still highly controversial whether such a rule can be considered as a genetic 'code', that is, as a semiotic system belonging to the genetic system.

Nucleotides do not know that A 'means' U. They simply react by substituting A with U.

We cannot say that nucleotides behave semiotically because we are unable to prove that they can refrain from interpreting or that they can choose alternative interpretations.

1.5. Stimulus-response

Let us consider two different cases: (i) I push a button and a bell tolls; (ii) I say rose and somebody answers "you mean red flower".

By pushing the button I implement a process that cannot but end with the tolling of the bell, while by uttering rose I implement a process that can end with these (or other) different responses: "you mean the past tense of to rise", "you are quoting a word used by Gertrude Stein", "And so what?", "I am not interested in what you are saying", or "buru buru aba boom!" (in Macomba dialect: "what kind of damned language are you speaking?").

The first phenomenon is based upon a stimulus-response mechanism, the second one requires a comparison betwen the received expression and a given sign-system, plus the decision of interpreting the expression.

A stimulus-response process is a diadic one: A provokes B and must be present in order to elicit B (equally present). I understand that the requisite of the co-presence is a very ambiguous one. A stimulus-response process is certainly a causal sequence, and we know many cause and effect sequences where the lapse of time between A and B is pretty consistent.

Let me say that I want to consider only cases in which such a lapse of time is so small that we can detect and record the physical presence of both A and B in the course of the same experiment. More complicated cases can produce a semiotic problem, but such a problem has nothing to do with the stimulus-response process, but rather with our ability to recognize it. In other words, I can decide that the present cancer of a patient X is due to a remote stimulation of his cells by certain radiations. Since I do not 'see' the past radiation, I infer its past existence by its apparent effect. I take the effect as the 'sign' or the symptom of its now absent cause. But if my conjecture was right, there was a regular cause-and-effect process, starting from the radiation and ending in the cellular response through an uninterrupted chain of physical states. The radiation is in some ways the absent Dynamical Object represented by the present cellular situation, but such a semiosic situation concerns my own competence, not the imaginary semiosic competence of the cell.

One can electrocute a man in California by sending an electric signal from a planet one billion light-years distant from the Earth, and we are hit by solar rays eight minutes after their emission, but in both cases we are entitled to consider A and B as co-present. On the contrary a semiosic process is always triadic: either A or B are absent and it is possible to see one as the sign of the other on the grounds of a third element C, call it the code, or the process of interpretation implemented through the recourse to the Code.

I beat your knee with a little hammer and you kick. Usually, except in pathological cases, it seems that you cannot refrain from kicking. Kicking becomes for you a blind necessity. The process is a diadic one. But now suppose that I beat you in 1980 and you kick in 1985. My problem is twofold: (i) is the beating A the cause of your belated kicking B? During this lapse of time, could you have decided freely whether to kick or not? I am unable to answer a similar one. I tell you "please kick" and then I wait. It remains unprejudiced whether you will kick or not (you can also respond by reproducing a new to-kick-or-not-to-kick Shakespearian situation).

One could say that the time-lapse between my order and your response is filled up with many intermediate steps and that my present problem is not so diifferent from the one concerning the link between a remote radiation and a present cancer. It is possible to object that even in the case of the radiation the cell could or could not have responded as it did. I am unable to make a decision on this very technical point. However I know that in human semiosic processes we have at our disposal more unquestionable evidence. If I asked ten different persons to kick I would certainly get ten different interpretations of my command. What is more relevant, I could also get my different interpretations of these ten interpretations, and a simple factorial calculus can tell you how many interpretations can be produced by my initial expression.

It is not necessary to contrast a high (human) behavior to a low (biological) one. It is sufficient to design two different abstract models: (i) a triadic model, where between A and B there is an unpredictable and potentially infinite series of Cs, and (ii) a diadic model where A provokes B without any mediation. C is a space of choice and of supposed indeterminancy, while the non-space between A and B is a space of blind necessity and of unavoidable determination. Many of our human affairs are ruled by model (ii). I shall have no difficulty in accepting the idea that many biological processes are ruled by model (i), provided it can be proved.

1.6 The C space.

Perhaps some day science will demonstrate that the C space is only a figment like ether, presupposed in order to fill up an 'empty' interval where deterministic phenomena - that escape our present knowledge - take

place. But until that moment we have to deal with C spaces. In any case, we know that in the C space a phenomenon takes place which is semiotically detectable: communicative contexts. Human beings are not emitting signs in the void, they are 'speaking' to and among other subjects that speak at their turn. In my semiotics a good sign-system also provides contextual selections. It takes the following format:
the expression X stands for a content that in $context_1$, will be y and in $context_2$ will be k.

I use to exemplify such a instructional semiotics by referring to the competence that allows us to interpret a stop signal at the crossing of two perpendicular roads:

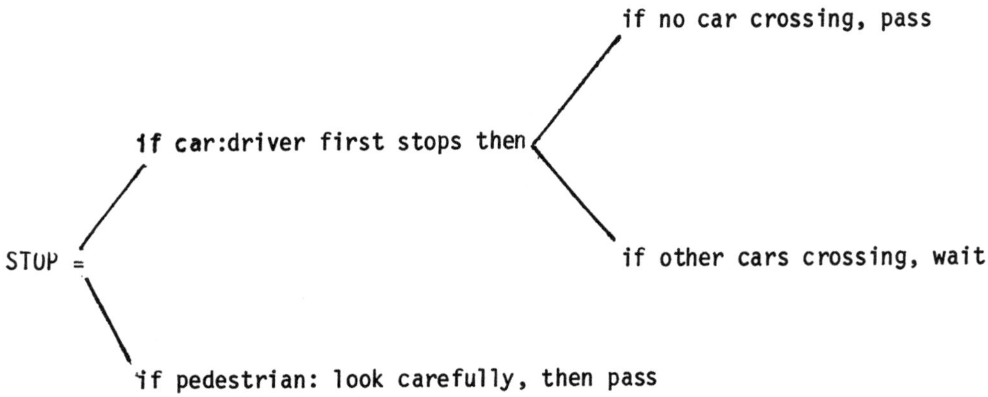

The model can be further complicated by considering the legal and physical consequences following the refusal to obey the instructions. In this case the C space in which the instructions are interpreted could also be characterized by the following features.
- the receiver can still suspect that the sender lies;
- the receiver does not know the code;
- the receiver understands the message and decides to disregard it.

However these features presuppose a conscious subject, and Celada remarks in his paper that in my presentation at the symposium I conceded that consciousness is not a prerequisite for semiosis.

I would like to elaborate a little more on this point, since it seems to me a possible meeting round for semioticians and immunologists.

2.1 Semiosis without consciousness.

Even a computer can be fed with my stop-instructions (see above) - provided it is endowed with a device allowing it to detect states of the external world such as the passage of cars. In this case one could object that the alleged triadic process becomes a diadic one where the stimulus-response chain is simply segmented into a series of binary choices, rigidly determined. We can say the same of any stimulus-response process: I beat your knee and (choice) if you are an a normal subject you kick, and if you are stoned you don't.

The basic instruction 'if-then-else' is simply a binary-structured stimulus-response process. Such a process could be made up with billions of necessary binary disjunctions and will still be a blind series of unavoidable choices.

But a normal home or personal computer is not a paramount example of Artificial Intelligence. Artificial Intelligence costs a little more than one or two thousand dollars.

Suppose we set up a system made up with three towers on the top of three hills, each three miles distant from the other. On Tower 1 we put a mechanical Device 1 which emits an electrical signal A when somebody enters the tower. On Tower 2 we put a Device 2 which lights up a lamp when a photocell detects the signal emitted by Device 1. Such a relationship is a necessary one; Device 2 detects the signal emitted by Device 1 and reacts in the only possible way. Then we put on Tower 3 a Device 3 which is instructed this way:

- when you see a lamp lighting on Tower 2 you shoot with your laser gun and destroy Tower 1 provided that:

(i) it is Monday

(ii) Reagan dislikes Regan

(iii) 'to dislike' is intended in the sense in which Othello disliked Desdemona

(iv) somebody told you something that you dislike

(v) you have previously made the above decisions and you have shot without succeeding in destroying Tower 1.

Everybody will agree in ascertaining that Device 3 is requested to make a difficult series of decisions according to a series of pretty ambiguous instructions, and that in order to understand them Device 3 must extrapolate from its past experiences. In fact Device 3 is requested to identify unheard-of contexts and to restructure its system of instructions (in other words, to produce a new code for every new situation). When Device 3 is able to do that we speak of Artificial Intelligence.

In order to succeed in its task Device 3 is requested to make some abductions.

2.2. Abduction

Abduction is an inferential process (otherwise called hypothesis) which is opposed to Deduction in so far as Deduction starts from a Rule, considers a Case of that Rule and automatically infers a necessary Result. A good instance of Deduction is:

(i) - every time A beats then B kicks

(ii) - but A beats

(iii) - then B kicks

Suppose now I do not know anything about that and I see B kicking. I feel astonished by that strange Result (iii). Since I know by previous experiences that usually one kicks when another one beats. I try to figure out a still unknown rule (i). If rule (i) held and if (iii) were the result of a case (ii), then (iii) would no longer be astonishing.

Remark that it is not necessary that I detect that frequently B kicks when A beats. This would be a case of Induction and nothing would guarantee my explanation (since it is enough to discover that B, at least once, kicks without being beaten, and my whole theory will collapse). In abduction, on the grounds of previous experiences, I guess that rule (i) could be a convenient solution in order to explain result (iii) and I try. Obviously my hypothesis must be tested in order to transform it into a law, but there are (in semiosis) many cases in which I am not looking for Universal Laws but for an Explanation able to disambiguate a single communicative event. You say rose and I do not know whether you are meaning a red flower or the past tense of to rise. Since however I know that you are a botanist, I bet upon the first possibility. If you react positively, I start going with our communicative interaction without wondering whether in the future you will be only compulsively concerned with flowers or not (there can be other cases in which you will be concerned with grammar). Naturally, in order to make a good abduction I must take into account the context. A context like I grew a rose is a paramount example of a disambiguating clue: according to English grammar, people are not supposed to grow past tenses (but who knows?)

To summarize: abduction is a typical procedure by which in semiosis one is able to make difficult decisions when following ambiguous instructions.

I am not saying that this is a nice and comfortable procedure. I am saying that it is the way we deal with semiosis at its higher levels.

The problem for immunologists is to decide whether they - when dealing with lymphocytes - are facing similar phenomena or not.

2.3 Recognition

Immunologists frequently use the word recognition.

I walk, I put my foot upon a hole in the ground and I stumble. Would I say that I 'recognize' the hole? I think that most of the 'steric' phenomena considered by genetics belong to this kind of mere stimulus-response process. How do the protagonists of the 'immunological system' recognize each other?

Suppose we teach Bongo, a chimp, to wear gloves and we give him several pairs of gloves of different sizes (morphologically indistinguishable from each other). Bongo will try them, and as far as a given pair fits his hands, he will take it as a good one. We would not say that Bongo 'recognizes' his gloves. He simply takes those that work.

Now suppose on the contrary that we teach Bongo to recognize his gloves by the following morphological features: they must be brown and marked with a golden star. Bongo learns on Monday that features of his personal gloves, sees on Tuesday a pair with the same features and recognizes them as his own (it is immaterial whether they fit his hands or not and whether he got wrong or not - in human terms to recognize means to believe that the X I see now is the same as the X I saw yesterday, and there can be wrong recognitions).

What happens when Bongo or Albert Einstein recognizes something?

Let us say that one is comparing both his actual perception X_1, and his memory of a past perception X_2, to an abstract type X. One switches from a presence (an actual perception) through a model, to an absence (a past perception). Recognition is always a triadic process. Recognition is always the comparison of two Tokens (one actual and the other remembered) to a Type. I see a pencil, I remember the pencil I used yesterday, and I decide that both are the same pencil because I compare them with my mental type X. The process is a triadic one because it is not necessary that all the features of the pencil I keep in my memory coincide with the features of the actually perceived pencil. Probably in the meanwhile somebody used it, and therefore decreased its length. My type pencil X is cylindrical, yellow, and marked with a star at one end. The length of the actually perceived pencil is irrelevant. It is only by this the mechanism that we are able to recognize somebody we met ten years ago even though his weight, the quantity of hair in his beard, the wrinkles on his face or the number of his teeth are changed. We recognize the X_1, object of my actual perception, as the same as the X_2, object of my past perception, if we have kept in our memory an abstract Type X that took into account only some pertinent features. Otherwise we would not be able to recognize today a certain John that we met twenty years ago when he was only in his twenties (and frequently we do not succeed in doing so, because in our memory and according to our emotional criteria "our" John was a man with a lot of black hair).

There is no general rule for establishing pertinence criteria. They depend on one's own needs. For a general who needs three soldiers to send against the enemy, to be recognized as a soldier few pertinent features are requested: supposedly, an average age and size, male sex, and a uniform. The general is not interested in the differences between John, Frank and Paul. For a lover who tries to recognize his/her partner after many years more pertinent features are needed (If I loved Sophia Loren I cannot mistake her with Rosalind Russell).

In semiosic processes the criteria for recognition change according to different contexts. Can immunologists say that the same happens with their cellular pets?

2.4 Models and metaphors.

If the answer to the above questions is negative, this does not mean that immunologists must avoid semiotic jargon. It only means that they can use semiotic models uniquely as models.

It has been frequently said that in science one should not use metaphors. But Max Black in his Models and Metaphors (Cornell Univ. Press, 1962) has persuasively distinguished between the linguistic use of metaphors and the scientific use of models. There can be both scale and analogue models. In both cases a model selects a set of pertinent features of the object it mirrors or maps. Scale models aim at reproducing the shape of the original object while analog models aim at reproducing only abstract structures, or systems of relations, of the original object. Moreover they reproduce the original object by another medium and by another web of relationships. Thus an electric circuit can be used as a model of a hydraulic phenomenon, or for a mathematical problem (and vice versa). A geographical map is a scale model of some elements of an analog model.

In this sense the functioning of a human language can be taken as a model for biological phenomena, or vice versa. Two conditions should hold at least:

(i) One must know that the map is not the territory.

(ii) The properties of the model must be better known than the properties of the object.

I do not know how much and to what an extent the second requirement is met by a semiotic approach to immunology. I do not know how much you know of immunological systems but unfortuantely I know how little semioticians know of semiosis and signification systems.

It may be that the second requirement is not as mandatory as it seems. Usually we believe that the more complex must be explained through the more simple. This way is called Reductionism. Following this way a semiotician can try to explain language by using the genetic code as a model. I do not trust so much such a way.

A fruitful alternative can be to explain the more simple through the more complex. I like this idea. It is daring but it can be productive. In this sense I will not discourage immunologists from using semiotic models.

However, at the beginning of my paper I said that perhaps immunology can help semiotics. Let me avoid the question of whether immunological phenomena are simpler than semiotic ones. After our symposium I realized that you are dealing with something that seems to be highly complicated.

What I learned from our meeting is that between the moment a lymphocyte meets an antigen and the moment it reacts, there is a space, and in this space there are some contextual choices to make.

My conclusion is that your C space seems simpler than our own. Nevertheless it still is a space.

Is a triadic process more characterized by the unpredictability of its C space or by the simple and dramatic existence of a space? If we accept the second answer (and I think that Giorgi Prodi will take this way) then the very existence of a semiosic phenomenon is guaranteed by the very existence of a C space and not by its complexity and unpredictability. Which means that in the depth of biological processes lie the elementary mechanisms from which semiosis springs.

What remains to be clarified is whether the width and the unpredictability of a C space represents or not the threshold between high and lower biological processes - or if the complexity of the C space is only another 'optical' effect due to the limits of our knowledge.

As you probably understand, such a question concerns the dramatic problem of the boundaries between Spirit and Matter, Culture and Nature.

Let me stop. I feel afraid.

A NONRESTRICTIVE SEMIOTICS OF THE IMMUNE SYSTEM

Patrizia Violi
Department of Communication
University of Bologna

I would like to start with a personal remark. When the opportunity comes to discuss semiotics with people from a different disciplinary area, I feel both hope and fear. I wonder, then, whether we semioticians are finally getting into the market, close to selling our theories and to showing that they are really productive; or whether too much work will be required out of us to make them really a little productive for others to use. At Il Ciocco, I tried to be neither too hopeful nor too fearful. Yet now, in putting down my ideas I see I can only sail between those two feelings. Is Semiotics relevant to Immunology? I could be very strict, very defensive in a way, and give an elucidation of Semiotics such that the answer would be a clear-cut no. Extant theories of Semiotics would be very helpful at that, because they are meant to give a paradigmatic definition of what Semiotics is. But I will rather be tentative and try to explore a non-paradigmatic case, which is interesting in itself for a semiotician, and which could fit the case. The case is indeed not clear cut, not paradigmatic. In order to frame it, I had to reconsider the most basic methodological tenets of my discipline. What follows is a description of the problems I met, together with an attempt to explore an alternative idea from which, for the moment, more than definite answers, new questions arise. Within a semiotic-linguistic tradition, one learns first of all the distinction between natural and non-natural meanings or signs. Such a distinction is based on a rather intuitive difference between cases such as 'smoke means fire', and cases such as 'with his words he meant such and such'.

Recognition tests have been devised which allow us to decide which kind of meaning we are dealing with. The most basic test is that in the natural cases of meaning there is a causal and necessary link between the sign and what is meant, while in the non-natural cases the link is arbitrary. Smoke is causally related to fire, while there is no causality between the word 'glass' and the object to which it refers (or the mental image we have of that object, if we prefer). It has been noted (Grice 1982) that the specification of the non-natural meaning can be comfortably achieved via the use of phrases in quotation marks, while this is not the case for natural meaning, In a sentence such as 'Those black clouds meant that it would rain', 'it would seem rather odd to say that those black clouds meant "It will rain": it does not look as if one can replace the <u>that</u>-clause here by a sentence in

NATO ASI Series, Vol. H23
The Semiotics of Cellular Communication in the Immune System
Edited by E. E. Sercarz et al.

quotation marks.' This test shows another important feature of language (and in general of any semiotic system), namely its reflexivity, the possiblity to use the elements of a given system in a metalinguistic way.

In more technical terms, the features that a semiotic system should be endowed with have been described by Hjelmslev as follows: biplanarity, commutation and non-conformity.

The first criterion is the most relevant, and it says that a semiotic system must be composed of two different levels (or planes): expression and content. The second and the third criteria set specific constraints on the relations between these two planes. In particular, commutation allows us to decide when a change on the expression level produces a change on the content level (as in sin vs sing); non-conformity says that elements of expression and elements of content are not related in a one-to-one mapping (i.e., single phonemes are not meaningful units).

Let us now turn to the immune system. According to the constraints we have just set, it is hard to see how the immune system could be approached as a semiotic system. There are two features of the immune system that are often taken as exhibiting some semiotic function: the productivity of the immune system and the communication, or more properly, the information transmission among cells.

In a paper published in 1985, Jerne addresses the first of these two features. We know that the number of different antibodies that the immune system of a single individual can produce is extremely high: it has been estimated that it exceeds 10 million. Such a 'repertoire' has been characterized as 'complete' (Coutinho 1984), which, according to Jerne 'means that the immune system can respond, by the formation of specific antibodies, to any molecule existing in the world, *including molecules that the system has never before encountered.*' (my italics)

The last phrase of the quotation is crucial for our discussion, since it describes a phenomenon that parallels very closely a basic feature of human language, i.e., productivity, the property which makes possible the construction and interpretation of new signals. What is peculiar to all language-systems is the capability of competent speakers to produce and understand indefinitely new sentences of their language that they have never heard before. Chomsky refers to this feature of natural language as 'open-endedness', and it seems rather attractive to compare the open-endedness of language to the completeness of the antibody repertoire as Jerne does. However, there is one important way in which the two systems differ. In the linguistic production, not only do we recognize a given form (for example the sounds of language) but we also 'grasp its meaning', i.e., we attach to its formal components an interpretation which is not *causally* linked to them. Does the immune system do something like that, or does it merely react to new molecules in the same way as, for example, the nervous system would react to a new stimulus never experienced before?

The fact that the combinatory power of the system is very high and complex should not deceive us. Combinatory power, in itself, is by no means evidence of a semiotic functioning. From this point of view, I believe the analogy with the grammar of natural language is misleading,. It is true that both language and the immune system have a grammar, in that they both are composed of a limited set of elements and general principles (rules) to combine them, but here the analogy ends. What makes natural language a semiotic system is the existence <u>at the same time</u> of a grammar <u>and</u> a semantics, not the mere capability of combining the elements of the system according to given rules. But it is precisely the existence of something like an autonomous semantics that seems problematic in the immune system.

Similar remarks are in order if we turn to the question of celullar communication, which is closely related to the recognition problem., Receptors on T and B cells recognize each other in such a way that they are able to interact with each other, finding 'appropriate partners'. It is in order to describe the link between recognition and pairing that immunologists talk of 'cellular communication'. Here again the crucial question is whether we are facing a real semiotic system (which would allow us to talk properly of a communicative process) or a mere, even if highly sophisticated, stimulus-response pattern (where to talk of communication is merely a metaphorical device).

Obviously, the answer to such a question depends upon the criteria we decide to use; as I have argued so far, the usual criteria of the traditional semiotic approach put too strong constraints on what can be defined as a semiotic phenomenon. In this way, they rule out the whole domain of natural meanings. From now on, I shall refer to such a position as the <u>Restrictive</u> view of Semiotics, and I will explore some of the possible features exhibited by a <u>Nonrestrictive</u> view.

According to a Restrictive view, there is a sharp boundary line between semiotic phenomena and presemiotic phenomena. In Eco's theory the 'lower threshold' of the semiotic domain coincides with 'physical information', the area of mere stimuli. The example used by Eco (salivation in Pavlov's dog) strongly suggests a mechanical and non significant nature for these phenomena; beneath such a treshold it is therefore hard to imagine any semiotic functioning.

While the upper threshold is no longer present because all cultural phenomena (such as tool production, kinship and exchange of goods) can be studied semiotically, the lower boundary is sharply set, and it excludes the entire area of human physicality from the domain of semiosis. It is not only the immune system that is ruled out (together with all biological systems) but also any kind of psycho-physical phenomenon.

The Nonrestrictive view I am here hypothesizing is less concerned with emphasizing the differences between the presemiotic and the semiotic than with looking at what they have in common. From such a perspective, the two domains are not completely unrelated, and their relation can be described as a continuum

where it is possible to find common basic structures, even if these structures assume different relevance according to the level of complexity of the system under consideration.
However, in order to make the assumption that the presemiotic meaning is somehow the ancestor of semiotic meaning, one needs different notions from the restrictive ones we listed above. I would suggest that there are two notions which can be of help in setting nonrestrictive, and weaker, criteria for semiotic relevance: intentionality and consequence.

Let us start with consequence. We have already seen that the notion of necessary cause is a crucial one in deciding whether we are facing a semiotic (and non-natural) system, or a mere mechanical response caused by a given stimulus. If it could be proved that not everything within the functioning of the immune system is deterministically caused, this would provide a very good starting point. Unfortunately, I know too little about the immune system to have any well founded ideas on this matter. I can only make some suggestions and leave any final conclusion to immunologists. If I have understood correctly, the immune system produces answers that are not completely predictable,. In the same vein, I believe, Celada has suggested that cells can 'wait' before reacting to antigens for more information which will enable them to 'disambiguate' their nature. Is this enough to claim that we have crossed the line of wholly mechanical reaction, even if we are not in a fully developed semiotic system yet? If this is the case - and this is a question immunologists, not semioticians, should answer - we could say that we are in an intermediate area between mechanical and semiotic phenomena, where neither the notion of necessary cause nor that of arbitrary rule fully applies. What we need here is a weaker notion, not strongly committed to any of the terms in the causal versus arbitrary opposition. Consequence, as defined by Grice, could very well be such a notion.

According to Grice, the root idea in the notion of meaning, which in one form or another applies to both natural and non-natural cases, is the notion of consequence. 'If x means that y, then this is equivalent to, or at least contains as a part of what it means, the claim that y is a consequence of x. That is, what the cases of natural and non-natural meaning have in common is that, on some interpretation of the notion of consequence, y's being the case is a consequence of x.' (Grice 1982)

With such a notion we are not saying anything about the nature of the process we are describing (causal or arbitrary), and by imposing a weaker constraint on our object, we leave open the possiblity of linking together natural and non-natural phenomena, presemiotic and semiotic processes. This is precisely what Grice tries to do, suggesting a possible description of how to derive non-natural meanings from natural meanings. He represents a case where non-natural meaning is descended from, and analogous to, natural meaning. His account is very close to a mythical description of the origins of language, a kind of evolutionary tale, and it is not intended to be of any historical or theoretical accuracy. However, it is a myth designed to 'exhibit the conceptual link between natural and non-natural meaning', and this is precisely the link we need in order to frame our question.

The notion of consequence is too coarse, however, since it covers very different phenomena, from linguistic expressions to mere mechanical reactions. Let us therefore introduce another constraint. Roughly speaking, there are two possible approaches to meaning. The first is based on the notion of code, an arbitrary rule which links together expressions and contents, words and meanings. I will call this meaning1.
The second approach appeals to the notion of 'intentionality'. Intentionality is different from intention, since it implies neither conscious motivation nor explicit purpose to achieve a given goal. Its roots may be traced to phenomenology, where intentionality refers to the idea of 'tending to' something and is linked to the idea of value.

The notion of biological value was introduced by René Thom in his semiotic theory of biological regulation. Referring to the forms animals can perceive, Thom distinguishes between formes saillantes (salient forms) and formes pregnantes (pregnant forms). 'Salient' forms are configurations of discontinuous elements, i.e., forms that are different from their background. Some of these salient forms are biologically meaningful, in that they have to be recognized by a given species in order for that species to survive. Such is the case, for example, of an animal recognizing something as a dangerous predator or as a possible quarry. Biologically 'salient' forms are called 'pregnant' forms and they have biological value.

Now, is such a distinction extendable to the immune system? Can we look at antigens as salient forms which are recognized as relevant to, or meaningful for, the system (so that a given antibody has to be produced as a response to them)? In other words, is it possible to say that biological value can be attributed to certain salient forms in such a way that they become meaningful? If the answer is yes, we could define such a process as a 'minimal domain of meaning'.

This is a qualitative picture of the matter, needless to say, and it cannot be reduced to a quantitative combination potential of elements. As Petitot (1986) puts it, 'pregnancy' is a kind of 'fluid' diffused in the world of salient forms which transforms that world into a meaningful environment. (Intentionality could be described as what makes this fluid 'flow' around.) I will refer to this 'minimal domain' of meaning, which can be found in such a meaningful environment, as meaning2.

I would like to suggest that by combining together the notion of consequence and that of meaning1, we have something very close to the definition of a restrictive semiotic system. Y is a consequence of x in that it is conventionally associated with x through a code (as in the case of language). On the other hand, if we combine consequence and meaning2, we obtain what I would like to define as a nonrestrictive semiotic system, where there are neither conventional rules nor consciousness, but only what we could describe as minimal conditions of meaning.

Can we tentatively assume that the immune system belongs to these nonrestrcitive semiotic systems? It is important to note that we are talking here of the global system, with its global rules, not of the single elements within the system. At the level of single cells we may very well have mere stimulus-response patterns. However, if it is true that the whole system is something more than the sum of its component parts and its local rules, it is at the global system level that our question should be more properly asked. The answer will depend upon the evaluation of the features of the immune system. A first point, as mentioned above, is the possibility of non-deterministic answers in the system.

A second feature which deserves attention is what immunologists refer to as 'internal images' of the system. Is this merely a metaphorical expression, or is there evidence for the existence of an intermediate level of representation, as is the case in 'high' cognitive systems? If I have understood correctly, each immune system has something like a memory of its own immunological history. Could such a memory be described as a component of that representation level?

Again, the answers to such questions are not within the realm of the non-immunologist (such as myself). As a semiotician, I can only suggest possible criteria for looking at the data, as I have tried to do with the non-restrictive or weak semiotic model. I do not know whether this can be of any use to immunologists. But I strongly believe that the problem immunologists are posing is of crucial concern for semioticians, in that it forces us to reconsider the question of the lower threshold of the semiotic domain and the relationship between the presemiotic and the semiotic.

The shift in level of description obtained by using less restricted constraints implies a much more relevant shift in the perspective we could assume on semiotics and meaning. Indeed, looking for what non-natural and natural meanings (or restrictive and nonrestrictive semiotic systems) have in common means looking for some 'deep' structure beneath both. The notions of value, 'pregnancy', and intentionality belong to this deep level. One could argue that the search for such deep universal constituents of meaning is very vague, and that, by using them, I am trying to explain something quite obscure with something even more obscure. This might well be the case. However, this is a risk we have to take if we want to move from the domain of clear, rational, well-defined semiotic meanings to the more obscure domain where these meanings originate, because what is at issue here is nothing less than the quest for origins. Asking whether biological structures exhibit elements of semiotic functioning is indeed only another way of asking the more basic, and frightening, question: 'Where does meaning come from?'

REFERENCES

Coutinho A. et al, 1984, Immunological Review, 79, 151.

Eco U., 1976 A Theory of Semiotics, Bloomington, Indiana University Press.

Grice P., 1982, 'Meaning Revisited', in Mutual Knowledge, Smith N.V.(ed.), New York, Academic Press.

Jerne N.K., 1985, 'The Generative Grammar of the Immune System', Science, 229.

Petitot J., 1986, 'Prégnance', in Sémiotique. Dictionnaire raisonné de la théorie du language, vol.2, Greimas A.J., Courtes J.(eds.), Paris, Hachette.

Thom R., 1972, Stabilité structurelle et morphogenèse, Paris, Ediscience.

Possible contribution of biosemiotics to the problem of communication among lymphocyts.

Thure von Uexküll
University of Ulm FRG

1.) Preliminary remarks about the concepts of system and sign.

a) The experience of systemic structures in research.

The reason for discussing semiotics at this meeting of immunologists is the fact of being faced with a strange situation: Immunologists are forced to use unusual expressions in order to describe their observations. Expressions like "memory", "recognition", "interpretation", "individuality", "reading", "inner picture", "self", "nonself", "iso-, idio-, allotype" or "killing" - to list only some of them - are unknown in physics and chemistry. Atoms and molecules have no self, memory, individuality, or inner pictures. They are not able to read, to recognise or to interpret anything and cannot be killed either.

This means that it is not possible to solve the problems described with these expressions by translating them into physical or chemical processes or explaining them in physical or chemical terms.

Nearly a hundred years ago a similar situation occurred to Christian von Ehrenfels, the founder of Gestalt-psychology (Kofka 1935). He reduced the problem to the formula:

"The whole is more than the sum of its elements."

Very soon it turned out that not only were psychologists confronted with this problem but also biologists (Köhler 1922). In the meantime this discovery has been rediscovered by system-theory. Here the phenomenon was termed "emergence" (Medawar and Medawar 1977, Popper 1982) to describe te "more" which a whole exhibits when compared to its elements or to the sum of its ele-

NATO ASI Series, Vol. H23
The Semiotics of Cellular Communication in the Immune System
Edited by E. E. Sercarz et al.

ments. This designation means, as soon as an integration of elements develops into a system it shows qualities and potentialities unknown at the level of its elements. Therefore it is not possible to explain the new qualities by reducing them to the qualities of the elements. If we try to do so, the qualities we want to explain, vanish.

This fact can be interpreted by the assumption that the activities of elements block each other in a manner that allows only cooperation which maintains the system or are in its "interest". A molecule as a part of cell components can not exhibit activities which would be possible outside the cell. Cells as components of a tissue or an organ can only divide and multiply themselves under close restrictions - an so on. The recognition of these facts means acknowledging different levels of complexity or integration, of which each needs its own language to describe its phenomena.

b) Semiotic consequences.

Problems of languages and descriptions are semiotic problems. Thus it becomes evident that problems of system theory are at the same time semiotic problems. This explaines the consideration in the discussion of immunologists. But this consideration confronts them with different types of semiotics:

1) with cultural semiotics, which place the primary emphasis on linguistics (Eco 1972). It examines the signs which serve in communication between human beings (including the communication between scientists);

2) with biological or bio-semiotics; it examines the signs which are exchanged between living systems such as cells, plants and animals. Some literature differentiates between cyto-, phyto- and zoo-semiotics (Krampen 1981, Sebeok 1978);

3) with "protosemiotics" (Prodi 1977); this name can be applied to an examination of quasi-semiotic phenomena in the interaction of atoms and molecules.

These distinctions reflect different levels of complexity or integration. It is necessary to keep this distinction in mind, because we are always concerned with all of them: protosemiotic signs (e.g. physical or chemical processes) serve as sign-carriers (carriers of information) in the exchange of biosemiotic signs between cells, plants and animals. These signs must be translated into cultural semiotic signs (terms of a scientific language) in order to understand and communicate the observations. This is the reason why all scientific statements have an irreducible anthropomorphic input.

The following contribution will restrict itself to problems of biosemiotic and touch the two other aereas only occasionally.

2.) Some remarks on biosemiotics.

a) Concept and methodology.

The methods, definitions and terminologies of biosemiotics serve the description of signs which are transmitted by physical, chemical, electrical and other carriers of information in the communication of living systems (e.g. gestures of animals (Sebeok 1969), songs of birds (Todt 1986), smells of insects (Schult 1986) and so on). The methods of research are composed of analysis of receptors and effectors as well as of the behaviour of living systems in the context of their biography and the situations in which signs are received and produced.

For the interpretation of data the concepts of "semiosis" and "meaning" are of central importance. In spite of continuing discussions about these concepts (Todt 1986, Tembrock 1984, Anderson et al. 1984, Sebeok 1976) we can start with a definition given by Peirce (CP 5.484) which differentiates between causality and semiosis:

> "All dynamical action, or action of brute force, physical or psychical, either takes place between two subjects (...) or at any rate is a resultant, of such actions between pairs. But by 'semiosis' I mean, in the contrary, an action, or influence, which is, or involves, a cooperation of t h r e e subjects, such as a sign, its object and its interpretant, this three-

relative influence not being in any way resolvable into actions between pairs. (...) My definition confers on anything that so acts the title of a 'sign'.

According to this definition we have to consider "causality" as a concept for an interpretation which links two phenomena (as cause and effect) by the idea of transportation of physical energy by means of linear steps (chains of cause and effect). "Semiosis" provides a concept for interpretation which links the phenomena by a meaning which is defined by an interpretant in a circular way.

This concept was introduced into biology with the model of the "functional circle" by Jakob von Uexküll in 1920. Its principle of feedback was formulated on a mathematical basis as "cybernetic circle" by Norbert Wiener in 1943. It interprets the activity of living systems as a control of their input by way of their output. According to the concept of semiosis this model interprets the relationship of living systems to their environment as interpretions of the meaning, which the environments have for the systems with respect to their "referent values" (nominal values). The information which the receptor (as sensor of a cybernetic circle) transmits to the system about the difference between the qualities of the environment and the referent value is equivalent to a sign. The behaviour of the system to the environment is equivalent to a signified answer and the referent value is equivalent to the interpretant. It can be interpreted as a symbol for the "self-quality" of a system. Thus the model explaines how living systems select a small sector of the environment as their "subjective reality" and reduce the complexity to the elements of importance for them.

Stimulus-response is not a linear chain of cause and effect, since it depends on the state of the system (its referent value), whether a process in its environment becomes a stimulus or not.We must distinguish between "sign-carrier" (a physical event) and its meaning as a sign, which the sign-carrier is given by the interpretant of the system.

The relationship between a living system and its environment can so be described as "assimilation" (Piaget 1950) to an accustomed referent value. The development of this relationship can be explained with the concept of "accommodation" (Piaget 1950). It describes a situation in which the receptor of a system provides an information which cannot be interpreted by an accustomed referent value (interpretant). In order to be able to assimilate the environment anew, the system has to change its referent value until the output (the behaviour) controles the input (the activity of the receptor) again. In biology we find many examples for both: assimilation of environment by stabel interpretants and accommodation with changing interpretants; an we find a development of receptors towards more and more refined instruments for selection and recognition of patterns in the environment.

In the concept of <u>semiosis as assimilation</u> we find the "three subjects" of Peirce:

- the interpretant, it correspond to the referent value of a cybernetic system,
- the sign, it is equivalent to the difference between the quality of environment an the referent value of the system, expressed by the reaction of the receptor,
- the object, it is represented by sections of the environment which match parts of the system (the organism) by complementing them to functiional units, i.e. the signified answer of the system.

Examples:

- The resistence of water (a section of the environment) points to the fins of of a fish, complementing them to the functional unit "swimming". The resistance is (in this context) the sign, water the signified (object); the concerned referent values act as interpretants.

  The same holds true with the resistance of air and the wings of birds, or the resistance of ground and the feet of animals or men (Uexküll, J.v.1936).

- The epitope of an antigen (as a portion of the environment of cells in the organism with the possibility of nearly endless different antigens) points (as a sign) to the idiotope of an antibody or a receptor of a B- or T-cell, complementing them in the function of recognizing (assimilating) and eliminating dangerous agents.

The concept of semiosis as accommodation with changing interpretants can help us to understand changes in behaviour as effects of changes in receptor-function.

Examples:

- Changes in saliva production and in gastric secretion of Pawlow's dogs as a result of conditioned stimuli.
- Proliferation of antigen-activated lymphocytes as reaction to IL2 only after production of IL1-induced IL2 receptors.

b) Advantages of biosemiotic descriptions.

Biosemiotic concepts can help us to understand the relationship between living systems and their environment, since signs reduce the complexity of environments to sections of biological significance. In this way they create the "milieu exterieur" of living systems.

As the organism is composed of subsystems which communicate by signs, we can imagine that also the "milieu interieur" of the body is created by signs (the milieu interieur of the body conforms with the milieu exterieur of its subsystems e.g. cells).

Signs which relate the organism and its environment belong to a "higher" level of complexity or integration as "endosemiotic" signs which relate cells and organs within the body to the milieu interieur and to each other. The connection between signs of different levels of integration becomes possible by a process which may be called "translation" because it links the meaning of signs which belong to different systems and different codes, comparable to the linking of the meaning of words of different languages.

Examples:

- Pawlow's conditioned reflexes are the result of translations by linking meanings between signs of two different levels of integration ("exosemiotic" signs which connect organism and environment with "endosemiotic" signs connecting organs and cells within the organism).
- Signs which maintain the homeostasis (steady state) within the immune system by providing communication between individual cells may be "translated" into signs which lead to an alteration of the system as a whole, i.e. the immune-defence-reaction as a response at a more complex level of integration.

Thus the model of "semiosios" gives a basis for rational definitions of terms like "recognition", "interpretation". "reading", "memory", "self" and "nonself" in the description of biological processes without having to imply consciousness: "Recognition", "interpretation" and "reading" can be defined as "assimilation" e.g. of an epitope by a specific receptor of a B- or T-lymphocyte; "memory" is equivalent to a result of "accommodation" (a change of interpretant = receptor function) which is maintained by the system; "self" is equivalent to "referent value" of a system or in accordance to it; "nonself" describes a difference to the referent value.

Literature references

Anderson,M.,J.Deely,M.Krampen,J.Ransdell,T.A.Sebeok,T.v.Uexküll: A semiotic perspective on the sciences: Steps toward a new paradigm. Semiotica 52-1/2 (1984) 7-47.

Eco,U.: Einführung in die Semiotik (1972) München.

Ehrenfels,Ch.v.: ct. Kofka,K.

Köhler,W.: Gestaltprobleme und Anfänge einer Gestalttheorie. Jahrb. ges. Physiolol. (1922)

Kofka,K.: Principles of Gestaltpsychology. (1935,1950) London.

Krampen,M.: Phytosemiotics. In: Deely,J.,B.Williams,F.E.Kruze (Eds) Frontiers in Semiotics, Bloomington: Indiana Univ. Press (1986) 83-95.

Medawar,P.B.,J.S.Medawar: The life science. Harper & Row (1977) New York, Hagerstown, San Francisco, London.

Peirce,C.S.: Collected papers (1931/35) Vol 1-6, Harvard Universtity Press.

Piaget,J.: La construction du reél chez l'enfant (1950) Neuchâtel (Swiss).

Popper,K.R.: Die Theorie der Emergenz und ihre Kritik. In: Popper K.R., J.C.Eccles: Das Ich und sein Gehirn (1982) München.

Prodi,G.: Le basi materiali della signalisatione (1977) Milano.

Schult,J.: Zeichenvermitteltes Verhalten bei Spinnen. Zeitschrift f. Semiotik (1986) 8,3 253-276.

Sebeok,T.A.: The sign and its masters (1978) Austin, London.

Sebeok,T.A.: Animal communication: Techniques of study and results. (1968) Bloomington, Indiana.

Tembrock,G.: Biokommunikation, (1975) Reinbeck b.Hamburg.

Todt,D.: Hinweis-Charakter und Mittler-Funktion von Verhalten. Zeitschrift f. Semiotik (1986) 8,3 183-132.

Uexküll,J.v.: Theoretische Biologie (1920) Berlin.
Uexküll,J.v.: A stroll through the worlds of animals and men. In: C.Schiller and K.S.Lashley (Eds) Instinctive behavior, International Universities Press, New York.

Uexküll,T.v.: Jakob von Uexküll's The theory of meaning. Semiotica (1982) Vo. 42-1.

Wiener,N.: Cybernetics or control and comminication in the animal and the machine. (1948, 1961) Mass. Inst. Technol.

A NOTE ON SOME BASIC CONCEPTS OF THE SIGN

Constance Jordan
Department of English and Comparative Literature
Columbia University
New York, N.Y. 10001

A conceptual basis for understanding meaning in language was established by Saussure's analysis of the word as a sign in his Cours de linguistique generale. His analysis has proved to be a point of departure for further studies in linguistics insofar as they are concerned with the signifying or semiotic function of language (rather than with problems that are principally philological, having to do with the historical development of particular languages). The contribution of Saussurian linguistics to epistemology has been enhanced by investigations in another discipline--philosophy--that have attempted to understand how words have meaning in ordinary language. This work originated with logical positivists in England and has since been pursued chiefly by English-speaking philosophers who are sometimes referred to as "ordinary language philosophers." The best known of their publications is probably J.L. Austin's How to do Things with Words. Put very simply, their investigations attempt to understand meaning in language by analyzing the word as a sign in the context of a given utterance.

There is attention to "context" in both Saussurian linguistics and in ordinary language philosophy but the meaning of context is not the same for both disciplines. Similarly, there is attention to "ambiguity" in both disciplines, but they do not conceive of ambiguity in the same way.

The Saussurian sign

It is important to understand that for Saussure the sign is made up of two elements: the signifier and the signified. Both elements inhere in the sign. A sign is never merely a signifier. The signifier element in the sign points to something outside or beyond itself, what is signified by the sign, but that pointing is done within the sign and abstractly as it were. The relation between signifier and signified in language as we commonly know it is arbitrary: the word as signifier as no real direct or observable connection to the word as signified. The word "tree" does not evoke in a reader or listener a notion of tree unless he already knows English and that the writer or speaker writing or speaking the word "tree" also knows English. On the other hand, the relation between signifier and signified in some sign systems,

NATO ASI Series, Vol. H23
The Semiotics of Cellular Communication in the Immune System
Edited by E.E. Sercarz et al.

and arguably in some languages, is not entirely arbitrary: the signifiers and signifieds of hieroglyphic signs are an obvious example of such non-arbitrary relations.

English

fig. 1

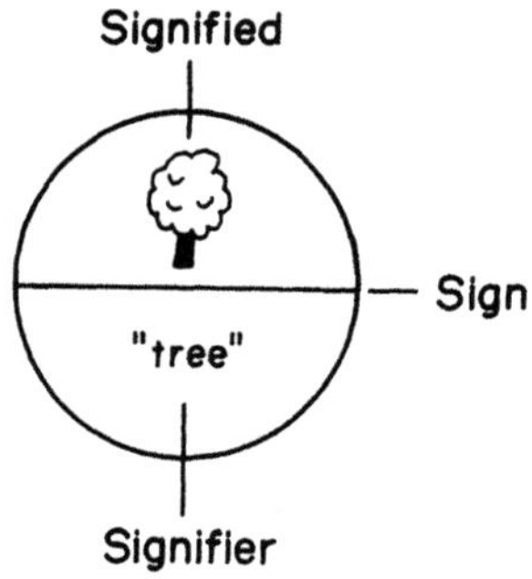

hieroglyphics

fig. 2

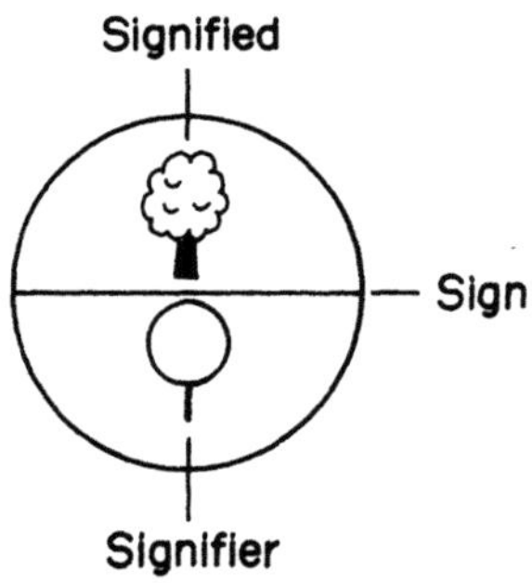

Saussurian "context":

Having established that the relation between signifier and signified is an arbitrary one, Saussure went on to postulate that meaning in language is not derived from a correspondence of the signified to a metalinguistic "reality" but rather from the relations between signs that are <u>generated within the language system to which they belong</u>. Meaning in language is therefore an auto-referential phenomenon; signs mean in the <u>context</u> of the system in which they appear. How the system is first established as meaningful is, I believe, a question Saussure does not answer.

Saussurian "ambiguity":

Some recent continental philosophers, and notably Jacques Derrida in De la grammatologie, who are interested in linguistics and meaning, especially in literary and philosophical texts rather than in ordinary speech, have chosen to focus further on the arbitrariness of the relation between Saussure's signifier and signified. They have stressed that in that arbitrariness there is an opening into a kind of dark ambiguity, an ambiguity that does not and cannot admit clarification. This has led them to postulate that the meaning of a particular word or statement is inherently elusive and that literary and philosophical words have meaning only in a tentative way that requires and indeed elicits further words. That is, any given word or words fail to signify meaning fully and unambiguously; this failure means that the process of signification is endless.

Some non-linguistic sign systems are practically free from the kind of ambiguity described above; that is, the relation between signifier and signified is so clear that the meaning of the sign is virtually unequivocal. A red light is a good example of such a sign. It has meaning in relation to a simple sign system containing three elements, red, green, and amber lights, and is understood in terms of a code that refers to the movement of cars and persons at intersections of streets and roads. A red light meaning stop almost seems as if it might not be a sign but a stimulus eliciting a response. But because it functions in a traffic code and therefore requires interpretation according to that code, it is a true sign and not a stimulus.

Ordinary language philosophers: signs and their referents

Meaning in language is also (and perhaps more usefully) analyzed as language is actually used to state ideas, issue directions, ask for favors, persuade to actions, and so forth--that is, as a means of communication between persons. At this point, the word as sign needs to be understood as it functions in a given conversational setting. A sender sends a sign to one who receives it; in addition to the significance the sign has by virtue of being a sign, e.g. its linguistic significance--which is, as Saussure saw it, always in a perspective of synchrony--it acquires a more particular meaning in that it is a sign sent in a certain act of communication. It has, in other words, a referent: as a sign it indicates or refers to some particular thing or notion that the sender wants to convey to the receiver. The referent is not the signified. The statement "cut that three down" can be diagramed to show the difference between the sign and its referent.

fig. 3

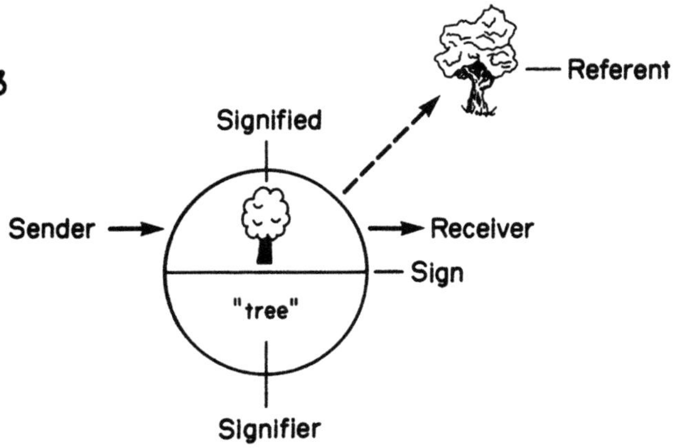

The statement is made by the sender knowing that if the receiver speaks English he will interpret it by grammatical rules as a command. If the sender raises his voice at the conclusion of the statement he knows that according to the conventions of spoken English it will be understood as a question: "cut that tree down?" If he stresses the word "that" he can expect that the statement will be understood to refer to a particular tree that needs to be distinguished from other : "cut that tree down!" These changes in tone of voice establish alternate meanings to the statement by contributing non-linguistic elements to its interpretation.

"Context" and "ambiguity" in ordinary language:

Signs that are picked up independently of the situation in which they are sent and received are notoriously susceptible to misinterpretation. Meaning finally depends on knowing a great deal not only about the meaning of the particular signs used in the act of communication but also about the situation of the sender (his motives, his linguistic training) as well as that of the receiver. In general: an awareness of the context of communication reduces the chances that what is communicated, the statement or sign, will be regarded as ambiguous.

There is, however, a final sense in which signs can acquire ambiguity (or better a suggestiveness) and that is if they are used figuratively, as, for example, in certain kinds of literary language. "Cut that tree down" might not, in the context of Shakespeare's Henriad, mean anything at all about using an ax against the trunk of a large botanical entity. It might mean something about the genealogical "tree" of the house of Lancaster. This kind of enlargement of meaning through using signs in a figurative sense is a consistent feature of poetic discourse. It is not to be confused with the ambiguity that colors statements made out of context.

Summary

Both the Saussurian sign and the ordinary language sign-in-context-of-act-of-communication require interpretation because they are substitutes for something that is not literally there. They mean what they mean by virtue of the fact that they are not the thing that they mean. "The map is not the territory." The stimulus/response interaction of the elements comprising the occasional immune response, in contrast to the interaction of elements within the system as system, does not appear to be susceptible of analysis in terms of a system using signs. No act of interpretation of a symbolic substitute occurs in this case. If, on the other hand, there are functions within the immune system that are carried on between elements in that system on a regular rather than occasional basis in which one thing stands or substitutes for another and is "read" as that other, then it is possible that the immune system may considered as a system using signs. In this regard, one might wish to pursue the answers to some obvious questions. What are the elements that function as true signs within the system? Do they, like linguistic or coded signs, need a particular context in order to be understood? Is the system, like a language system, auto-referential? Are these signs ever ambiguous? What factors, if any, contribute to their ambiguity, and does the system have a way of clarifying ambiguity as part of its self-maintenance?

INTRODUCTION TO TERMS RELEVANT TO IMMUNOLOGICAL RECOGNITION

Kristian Hannestad
Institute of Medical Biology, School of Medicine
University of Tromsø, Norway

This chapter is intended to orient non-specialists

1. B-epitopes and B-paratopes

When a micro-organism, for example influenza virus or group A streptococci, invades an individual, it usually elicits a specific antibody (Ab) response and secreted Abs establish a firm but reversible union with the micro-organism. Ab responses can also be elicited artificially by vaccination. The molecule that binds Abs is called antigen. An individual Ab binds to a specific area of that antigen which elicited the Ab and the association of the two is known as an antigen-Ab complex. In the case of protein antigens, amino acids in this area establish chemical bonds with amino acids of the Ab. The contacts between antigen and Ab is called the antigen-Ab interface where the contact amino acids of the antigen are collectively referred to as the epitope and those of the Ab are known as the Ab combining site or paratope. Since Abs are made by bone marrow derived (B) lymphocytes, epitopes defined by Abs are called B-epitopes. The tight fit of the antigen-Ab interface resembles the complementarity of a "lock-and-key" model (1).

An Ab displays a high degree of selectivity for a given epitope, that is, it binds firmly to antigens which express this epitope and does not combine with the vast majority of other antigens. This is known as immunologic specificity and is the basis for using Abs as analytic reagents, for example to identify drugs, hormones and blood groups. An Ab is usually named after its specificity; thus an Ab that recognizes sugars characteristic of blood group A is designated as anti-blood group A.

NATO ASI Series, Vol. H23
The Semiotics of Cellular Communication in the Immune System
Edited by E.E. Sercarz et al.

2. Antibodies are detector and effector molecules

Abs belong to a class of proteins called immunoglobulins (Igs); all Igs share a basic structure (1, 2). An IgG Ab molecule (the most abundant class of Ab in the blood) is shaped like a Y and binding to antigen (epitope) occurs at the tips of the limbs of the Y. Since for a given Ab-molecule each limb of the Y is chemically identical to the other, an Ab bears two identical paratopes both of which can attach to an antigen bearing at least two complementary epitopes. Abs confer new properties onto antigens: when the Y is bound, on the surface of a micro-orga-nism say, its stem is a sign recognized by specialized cells capable of engulfing and destroying the parasite. The stronger the union of antigen and Ab the more stable is the display of this sign. Specifically bound Abs represent molecular links between the foreign parasite and the non-specific defense (effector) cells. This function of the antigen - Ab complex is reflected in terms like "protecting" or "neutralizing" epitopes. Abs may also be regulatory molecules and possibly useful as vaccines (se section 7).

3. The clonal repertoire of B lymphocytes

Abs are embedded in the membrane on the surface of B lymphocytes where they serve as receptors for antigen. All paratopes of a given B lymphocyte are chemically identical, and lymphocytes that bear identical paratopes represent a clone. The Abs produced by a clone of B lymphocytes are called monoclonal; monoclonal Abs represent a chemically homogeneous population of paratopes. The number of structurally unique paratopes of a mammalian immune system is extremely high, in the order of 10^9 (Ref 1); this is known as the structural Ab diversity. When different Abs are compared the differences between them are located near the tips of the limbs of the Y; this region is called the variable (V) region. The remainder of the molecule is relatively invariant. The number of unique epitopes recognized by this B lymphocyte clonal repertoire represents the functional diversity. An antiserum raised in a animal against a foreign serum protein contains a large number of different Abs with specificity for the protein; thus, the specific Abs of an antiserum are nearly always polyclonal. Monoclonal Abs are constructed by fusing a single Ab-producing B lymphocyte with a non-Ab producing malignant lymphocyte capable of unlimited growth in cell culture.

4. Self/non-self discrimination and immunological tolerance

Given the enormous size of the functional Ab repertoire, a protein antigen has the potential to display an indefinite number of distinct epitopes to this repertoire. Virtually any part of the protein surface may express an epitope (2). The exception to this rule is antigens that are a natural part of the host of an immune system; these antigens are called auto-antigens or "self". It is usually hard to elicit Abs against autoantigens. This is known as immunological tolerance which is a fundamental property of the immune system. If one immunizes a rabbit with a protein homologous to one of its own, for example cytochrome c from mice, the Ab response is directed primarily to those regions of the mouse protein that differ from the rabbit. However, the tolerance is not absolute; small amounts of Abs with specificity for host (rabbit) protein may also be detected (cited in Ref. 3). The mechanisms behind immunological tolerance are not known. One possibility is that the clonal repertoire has been purged of clones with substantial affinity for epitopes of autoantigens; another is that such clones are actively suppressed. Thus, tolerance limits the functional repertoire of B cells.

5. The B-epitope concept

Clones that have bound antigen via their surface bound paratopes become targets for the powerful regulatory influence of helper and suppressor thymus dependent T lymphocytes. By complex processes certain B-cell clones are selected by antigen and regulatory T cells to divide and develop into high-rate Ab-secreting cells. The paratope of the Ab secreted by a given clone is identical to the surface bound receptor paratope of that clone. The epitope-paratope relationship has been studied in great detail with an Ab to lysozyme, a protein found in large amounts in eggwhite. Sixteen amino acids of lysozyme (the epitope) made close contracts with 17 amino acids of the Ab (the paratope) (1). A B-epitope can only be thought of in the context of an Ab, or stated otherwise, an epitope exists only insofar as it is relatable to an Ab (3). According to this concept the proposition "a large number of different Abs will react with a single epitope" is likely to be wrong because an epitope is not an inherent property of a molecule as is its molecular weight, absorption spectrum or amino acid sequence. Also likely to be incorrect is to refer to a B-epitope as immunogenic because it is improbable, in view of the vast repertoire, that a given antigen would regularly select the same clone in different individuals. Different Abs probably define distinct epitopes although the distinctions may be subtle (see below). A B-epitope should therefore be given the designation of the monoclonal Ab that defines it.

Certain limited areas of an antigen may be more frequent targets of an Ab-response than others; they are called immunodominant regions. Two (or more) Abs which compete with each other for binding to a protein may recognize two epitopes that occupy overlapping areas; such epitopes are said to be topographically related. A collection of topographically related epitopes is referred to as an epitope cluster. Epitopes defined by Abs elicited by intact globular proteins are often destroyed when the antigen is unfolded (denatured) indicating that they are assembled from two or more distant parts of the unfolded (linear) polypeptide chain. Epitopes of this kind are termed discontinuous or assembled. By contrast, immunization with a synthetic peptide representing a linear sequence of 10-20 amino acids of a given protein may elicit Abs that cross-react with the native protein; such Abs define continuous (segmental) epitopes.

6. Cross-reactions

The antigen which elicited an Ab is called the homologous antigen. An Ab that reacts with epitopes borne by antigens other than the homologous is said to cross-react. The work of Landsteiner indicated that all Abs have the potential to cross-react, which has been amply confirmed by subsequent observations. A monoclonal Ab, therefore, defines a set of epitopes, designated as serologically related epitopes (the specific reactions of Abs are usually called serologic reactions). This phenomenon is known as the degeneracy of Ab specificity, which implies that biochemical methods are required to verify the identity of an antigen; serological identification is not completely reliable. Since there exists countless different antigens in the environment it is often difficult to find the few that bear a related epitope.

Some of the epitopes of the serologically related set resemble the homologous antigen chemically as, for example, when a monoclonal Ab elicited by a given influenza virus hemagglutinin molecule combines with reduced strength to a mutated variant of the hemagglutinin. Since the two hemagglutinin variants differ from each other by very few amino acids, the epitopes recognized by this monoclonal Ab are likely to share many amino acids. Chemical similarity is the molecular basis of conventional cross-reactions known since the work of Landsteiner. For some protein antigens there exists a panel of variants that differ from each other by only one or very few amino acids. Such panels can be exploited to define the cross-reactivity patterns (also called the fine specificity) of monoclonal Abs. Thus, if we have generated serveral monoclonal Abs by immunization with one member of the panel we can select for study those (three say) that compete with each other for binding, indicating that they recognize the same antigenic region. We may want to know whether they possibly recognize the same epitope. To this end we determine the fine specificity of each of the three Abs for the panel of variants. If the three Abs can be distinguished by their cross-reactivity patterns we may conclude that they recognize different epitopes.

Unanticipated cross-reactions have been observed with monoclonal Abs isolated from patients with systemic lupus erythematosus or lymphomas. For example, an Ab has been shown to react specifically with structurally unrelated substances such as nitrophenyl and nitronaphthyl groups, denatured DNA, RNA, heparin, dextran sulphate, polystyrene sulphonic acid, a ribitol teichoic acid and a bacterial ribose-phosphate polymer (4). The molecular basis of these strange cross-reactivities remains unknown. Similarity of spacings of negatively charged groups would seem to be one structural feature that may at least partly account for these "polyspecific" antigen-Ab interactions. Whatever the explanation, it is clear that serological relatedness does not necessarily imply conventional chemical similarity. Sometimes the term "molecular mimicry" is used to describe this phenomenon. It is not known whether Abs that exhibit polyspecificity represent unusual clones.

7. Internal images - idiotopic mimicry

Since Abs are protein molecules they also express B-epitopes. Epitopes located at the tips (V-regions) of the Y shaped Ab molecule are called idiotopes. The repertoire of idiotopes is at least as large as the structural Ab diversity. Therefore, some Ab variants in the repertoire may by chance bear an idiotope that resembles more or less closely an epitope of a conventional antigen. For example, an anti-poliovirus Ab will cross-react with an Ab, if the latter Ab bears an idiotope which mimics a polio-epitope. This is called idiotopic mimicry which is an interesting reflection of the countless Ab-variants of an immune system as well as of the degeneracy of Ab specificity. Such idiotopes were termed "internal images" by Niels Jerne (5). Since an internal image is likely to be only partially similar to the conventional epitope it mimics, it was called "related epitope" by Nisonoff and Lamoyi (6). The quality of relatedness can be exploited. For example, if the idiotope mimics a neutralizing or protecting epitope of an infectious agent, then the Ab carrier of the idiotope may be used instead of the agent as a vaccine. This is a very recent application of Abs. A problem with idiotope-based vaccines may be that immunity as a rule depends on activation of both B- and T-lymphocyte clones; a vaccine should therefore contain both B- and T-epitopes derived from the agent in order to obtain a natural boost upon infection (7). Since a B-epitope is usually not recognized by T-lymphocytes it may be very difficult to achieve this goal with an idiotope-based vaccine.

Inasmuch as the paratope of one clone may recognize an idiotope of another (see above), the possibility exists that clones within an immune system may interact via their receptors for antigen. This is called connectivity within the repertoire and its possible regulatory significance is further elaborated by others in this volume.

8. T-epitopes, restricted recognition and allorecognition

Thymus-dependent (T) lymphocytes regulate the onset, magnitude and duration of immune responses by secreting lymphokines, chemical signals detected by various cells of the immune system. Secondly, T-lymphocytes can more directly engage in destruction of invaders by killing virus-infected, foreign (grafts) and cancer cells. T-lymphocytes that augment Ab-responses are called T helper cells; those that down-regulate immune responses, T suppressor cells; and those that kill, T cytotoxic cells.

T cells, unlike B cells, are unable to recognize free antigen molecules. Furthermore, T lymphocytes are only activated by antigen on the surface of cells, not in the fluid phase. To be recognized by T-lymphocytes there is evidence that protein antigens must by processed, which involves uptake by a cell, intracellular degradation and transport back to the surface where fragments of antigen may accumulate in the lipid membrane. Here the processed antigen is thought to engage in transient interactions with a special kind of protein molecules called immune response associated (Ia). All cells that express Ia have the potential to present antigen to and activate T helper cells and are called antigen presenting cells (APC).

The T-lymphocyte's antigen receptor (TAR) is embedded in the surface membrane and consists of two covalently linked polypeptide chains (α and β) whose basic structure resembles the limbs of the Y of the Ab molecule (8) although their folding pattern is unknown. The T-cell clonal repertoire, like its B cell counterpart, is extremely large, but is limited by immunological tolerance (see section 4). Current opinion on the molecular basis of antigen recognition by T cells is more tentative than for Abs. There is evidence that the TAR of T helper cells bind a complex of native Ia plus processed antigen in the lipid membrane of APC's. This complex is called the stimulation complex for T cells. A T-epitope expressed by the stimulation complex is believed to be assembled from amino acids some of which derive from the Ia and others from the peptide fragment (9). Studies of T clones with specificity for small peptides of 12-15 amino acids have indicated that a peptide of this size can contribute to the formation of several T-epitopes. This was shown by the distinct cross-reactivity profiles exhibited by the T cell clones for peptide analogues that differ by single amino acids (see section 6 above) or for peptides that differ in length. Thus a short peptide is precisely defined with respect to molecular weight and amino acid sequence but not with respect to the epitopes it has the potential to express. A T-epitope should therefore be given the designation of the cloned T-cell line that defines it.

There exists many Ia-variants in a species; the variants resemble each other, they only differ by one or a few amino acids. Those variants with which T cells coexist in a given individual are said to be syngeneic; all the other variants of Ia that exist in a species would be foreign to this individual and are called allogeneic. When T cells of this individual respond to a protein antigen, the Ia element of the stimulation complex is therefore a syngeneic variant. T cells thus stimulated cross-react poorly or not at all with the same peptide in the context of allogeneic Ia. Hence T cell recognition of protein antigens is restricted by Ia. The molecular basis for restricted recognition is not clear. One possibility is that allogeneic Ia lack critical amino acids necessary for formation of T-epitopes of the original stimulation complex. Another is that allogeneic Ia and processed antigen fail to form complexes.

The Ia variant of the stimulation complex exerts a strong influence on which particular region of a protein antigen is recognized by T cells. This phenomenon is called determinant selection and is an example of how an inherited host factor (Ia) directs the T cell response to a limited region of the protein antigen which thus becomes immunodominant in that individual. The molecular basis of determinant selection is not known; the phenomenon may reflect that a given Ia has a single distinct binding site which binds a limited peptide region of a protein antigen more firmly than other regions and therefore preferentially form stimulation complexes with that region (10,11).

A number of T cell clones with specificity for a given protein antigen have been observed to cross-react with APC expressing allogeneic Ia in the absence of that antigen. This indicates that allogeneic Ia sometimes mimic T-epitopes expressed by a complex of syngeneic Ia plus peptide antigen. An extraordinarily large part of the T helper cell clonal repertoire appears capable of recognition of allogeneic Ia. This is called allorecognition which unlike restricted recognition proceeds in the absence of processed peptide antigens. Allorecognition plays an essential role in the rejection of organ transplants (allografts). Ia are therefore also called histocompatibility molecules. There is evidence that T-epitopes involved in allorecognition are assembled from two or more distant part of the unfolded Ia. It will therefore be interesting to learn to what extent B- and T-epitopes expressed by allogeneic Ia resemble each other.

Literature Cited

1. Amit, A., Mariuzza, R., Phillips, S. and Poljak, R., Three-dimensional structure of an antigen-antibody complex at 2.8 Å resolution. Science (USA) 233, 747-753 (1986).

2. Leder, P., The genetics of antibody diversity. Scientific American, May 1982.

3. Benjamin, D.C. et al., The antigenic structure of proteins: A reappraisal. Ann. Rev. Immunol. 2, 67-101 (1984).

4. Tolleshaug, H. and Hannestad, K., Binding of ligands to a monoclonal IgM macroglobulin with multiple specificity. Immunochemistry 12, 173-182 (1975).

5. Jerne, N.K.. Towards a network theory of the immune system. Ann. Immunol. (Paris) 125C, 373-389 (1974).

6. Nisonoff, A. and Lamoyi, E., Implications of the presence of an internal image of the antigen in anti-idiotypic antibodies: possible application to vaccine production. Clin. Immunol. Immunopathol. 21, 397-406 (1981).

7. Berzofsky, J.A., Intrinsic and extrinsic factors in protein antigenic structure. Science (USA) 229, 932-940 (1985).

8. Marrack, P. and Kappler, J., The T cell and its receptor. Scientific American, February 1986.

9. Schwartz, R.H., T lymphocyte recognition of antigen in association with gene products of the major histocompatibility complex. Ann. Rev. Immunol. 3, 237-262 (1985).

10. Babbitt, B., Allen, P., Matsueda, G., Haber, E. and Unanue, E., Binding of immunogenic peptides to Ia histocompatibility molecules. Nature 317, 359-361 (1985).

11. Guillet, J.-G., Lai, M-Z., Briner, T., Smith, J. and Gefter, M., The interaction of peptide antigens and class II major histocompatibility complexes antigens as studied by T cell activation. Nature 324, 260-262 (1986).

DO YOU RECOGNIZE ME?

SIGNS AND CODES IN IMMUNOLOGY

G. Prodi
Istituto di Cancerologia
Università degli Studi di Bologna
Viale Filopanti 22 - 40126 BOLOGNA, ITALY

1) - Semiotics and immunology

Immunological reaction consists in deciphering molecular structures (antigens, antigenic epitopes). This is demonstrated by the response itself (production of antibodies or antibody-equivalents). The response is specific, i.e. is directed only towards the single antigenic structure by which it has been provoked. Therefore the immune system performs an "interpretation" of the material objects that it can explore, and is triggered by some of them: toward these it builds its response. In a very general perspective, antigens are "meaningful things" for an immune system, because of their capacity to trigger the system in a specific way. They are therefore "signs" which the immune system is able to interpret. Immunocompetence is a "system of interpretation", which works on the basis of a particular physiology and with particular rules. It is a "code" of interpretation. The property to be a sign only exists in relation to a system of interpretation: a molecule is an antigen only in relation to an immune system. An "absolute sign" does not exist as far as an "absolute antigen" does not exist. It is accordingly justified to attempt to verify if the concepts of semiotics (as they emerge from the study of linguistics and other human sciences) can be applied to immunology. Caution and care must be used, because very distant cultural domains are involved, and empty metaphors can easily be produced: in both the cases it is a matter of things which are identifiable as signs by a deciphering system, in relation to which they are properly signs.

2) - Systems of deciphering

Molecular structures are identified by the immune system as provided with a definite special shape resulting from the sequence of their

NATO ASI Series, Vol. H23
The Semiotics of Cellular Communication in the Immune System
Edited by E. E. Sercarz et al.

constituents. This brings about a sharp discrimination between the "self", against which the reaction does not occur, and the "not self". The laws that regulate the reaction must therefore also explain its absence.

Among not-self molecules, definite structural traits allow the distinction between one antigen and the others. The single epitope is confronted with (introduced in) a system of interpretation by which it is manipulated and explored. The single immune system is confronted with (being internal to) the epitopic universe which it is capable of interpreting. This interpretation can be in some extent ambiguous (cross-reactions among different antigens, as they were synonyms; autoimmune reactions; tolerance against not-self molecules), and this fact causes pathological conditions, or can be used in prevention or therapy, Namely, it is possible to manipulate the system, if its rules are known.

The attitude of the immune system to read the world is comparable to that of the nervous system, though in a very different way and on very different objects. The nervous system is devoted to the discrimination of inputs from the environment (inside and outside) in term of localisation and characterisation of the emitters, and is based on cooperating elements working in a fixed and invariant position (neurons, synapses). The immune system works by discriminating molecules present inside the body, through elements which cooperate, move, and collide in a free medium. As far as the nervous system is concerned, during the evolution of man such a complexity was attained, that allowed the formation of language and the interpretation of reality through a symbolic medium (semiotics as usually considered). It is matter of higher order: in a linguistic system a sign (e.g. a word) is "triadic" and its connection with reality is mediated by the body of the language itself. What is the relationship between these two signs (a word and an antigen), so distant in the domain of reality?

The problem is that of the semiotic threshold. When, during evolution, can we attribute the character of sign to a matter of fact (thing or event)? When does it raise in nature the occurrence of a thing which is "meaningful" for another thing? The semiotician usually fixes the threshold at the level of man, saying that sign is something which stands for something else in the frame of a language.

We are searching for the natural ancestors of this very recent (human) situation. In our opinion the threshold has to be placed much behind.

3) - Biology as natural semiotics

We think that the threshold for "sign" is situated at the very beginning of the biological domain, characterizing its origin and its basic structure. The pre-biological universe is consituted by standard and uniform situations, both of chaotic-statistic type (plasma, gases and liquids) and of repetitive type (crystals). Any condition of selectivity or specicifity is absent. Life begins when to such uniform world, conditions of selectivity are superimposed or, better, when conditions of selectivity are generated from the conditions of uniformity. This status is possible because of the production of molecular structures endowed with sufficient complexity, proper to the chemistry of carbon. An enzyme, wich can be considered the simplest example of this status, selects its substrate among a number of meaningless molecules with which it can collide: it reacts and forms a complex with only its molecules partner. This substrate is a sign for the enzyme (for its enzyme). The enzyme explores reality and finds what corrensponds to its own shape: it is a lock which searches and finds its proper key. In philosophical terms, an enzyme is a reader that "categorizes" reality determining the ensamble of all the molecules which can factually react with it. The term category is used here both in a Kantian sense (a "point of view" of a reader who gives order to the world) and an Aristotelian sense (the objective ensemble of things presenting a given character). This semiotics (or proto-semiotics) is the basic feature of the whole biological organisation (protein syntesis, metabolism, hormone activity, transmission of nervous impulse, and so on). In all these cases a specific relationship between a reader and its sign is clearly established.

Often a molecule is at the same time sign and reader. An organism is constituted by (results from) a high number of these correspondances. Biology is, in itself and in all its aspects, a natural semiotics. (This topic is treated in much more detail in the works quoted in bibliography). This "protosemiotics" constitutes also the ground of the immune reactions.

If a matter of fact is a sign, the reader for whom such a matter of fact is a sign has a proper code of interpretation, and only on the base of this code the matter of fact is a sign. If a molecular structure is interpreted by the immune system as an epitope (and is therefore a sign for the system) the question is: how is the code constituted, how does it work to interpret its reality. The mechanism must be based on the general well established reactions of deciphering (DNA coding, protein syntheses and so on). Then the question can become the following: how these reactions are combined to produce an "immunological interpretation". A novelty, in term of natural history, always is obtained as a more complex relation among ancient and well working mechanisms. Also the great novelty represented by the immune system follows this rule (the language too, in our opinion).

4) - <u>Semiotic repertoire</u>

The proof that a matter of fact is a sign for a code is the response of the system itself. The proof that a molecule is an antigen for the immune system is the production of a complementary specific molecule (an antimolecule, as a free antibody or as a membrane-associated protein), or can be also a specific interference on this production. The main point is the mechanisms which links the two terms: how the antigen is connected with the synthesis of its anti-structure.

The instructive theory mantains that the antigenic epitope has a primary function in the process, because, in some way, tha anti-structure is built on it: the antigen constitutes the mold for the production of its antibody. This does not exist before the antigen, both as molecule and information, and is constructed <u>after</u> and <u>on</u> the antigen.

The system has an endless capability to face all the possible epitopic structures.

Such a theory is impossible. Antibodies are proteins, and as all other proteins, are produced from an existing codification represented by DNA sequences. An instructive theory should be based on a totally unknown mechanism of protein synthesis.

The second possibility is that the genetic code contains a certain finite numbers of sequences, capable to rule the synthesis of the anti-structures for all the epitopes. In this case the function af the antigen is less important (through ever necessary), because it simply triggers the work of a preexisting machinery: it selects the single response from the repertoire of all the possible response. This is the ground of the selective theories. The system of interpretation (repertoire and rules) is an a priori of the interpretation of the single epitope: it is properly a code (in the general sense of a systematic modality of interpretation, in the same way that a language is a code).

This mechanism, only supposed till few years ago, is nowadays well known. In each cell of the body all the DNA sequences are present for the anti-epitopic structures, obviously in a finite number, but covering all the possible shapes of antigens. They are placed in some definite positions in chromosomes (for instance in chromosomes 14, 2, and 22 as far as heavy and light chains of man are concerned). In a definite set of lymphocytes these DNA sequences can be arranged by translation and joining with a random mechanism, so that diversity is generated starting from the same repertoire: in each cell a given combination of DNA pieces is reached, different in principle from any other. So from a repertoire of different sequences a repertoire of different cells is obtained: the semantic dictionary of the immunological code is distributed among a high number of cells, which are in this way different from each other and genetically specific. The core of the system is therefore an heterogeneous population of cells which as a whole can read all the epitopes, but only one of them (in principle) can read a single epitope (of course it is not completely true, because the same combination has a certain probability to be reached in different cells). The mechanism of VDJ and C joining is one of the most interesting results in molecular biology.

Therefore the system contains an internal repertoire (as codification of complementary structures) of the epitopic universe: it is an a priori mechanism in relation to the events it can decipher, into which it can fall. In this sense it is not an exception, because all the biological entities are a priori in relation to their environment, through a posteriori products of natural history.

In this sense we can say (using with cautions these expression) that antigens are the "totally alien" which is introduced into the domain of the language (of the known), being analyzed by a categorical system (made part of a process of interpretation, and, in this sense, "razionalized" by the system).

5) - From the repertoire to the syntax

The repertoire of the anti-epitopic DNA sequences constitutes the semantic component of the system. As in all the languages, also a syntactic component is necessary, which establishes the rules for the use of the repertoire: how a definite proposition can be obtained using the words of the dictionary.

The immunological repertoire is, per se, a pure potentiality, that must be transformed in an actual process to obtain the product of the reaction. Syntax means rules, and rules means essentially a restriction of possibilities. In fact a syntactic scheme in a language forbids most of the combinations among the words of a dictionary, and allows only few of them: it restricts drastically the high number of random combinations of the words resulting from a kind of thermodynamic collision, and establishes the permitted structures.

The "meaning" resides precisely in this process: from an indefinite casuality to an ordered situation, interpretable also in terms of the theory of information and of the theory of communication. It is matter of a "channelling" of the semantic entities into the definite pathway of the syntactic rules.

To use the repertoire, the first step for an antigen is to be handled in a proper manner by macrophages, to be exposed on the surface of a proper cell, and then some fixed processes among specialized cells can take place.

This is the channelling we are speaking about: each step of this represents a structure-antistructure selection trough specific receptors, both between two cells and between a lymphokine and a cell. Only at the end of the chain is the antibody forming cell triggered to clonal amplification and anti-epitope production. Also after this process, some other chains of molecular events must take place, to modulate the amount of the products formed. Many aspects of this picture (many rules of this grammar) are nowadays known, appearing as particular cases of the general "protosemiotics" we are speaking about. Cooperation between T and macrophages, between T and B, between T and T, the action of the different interleukins and so on are all chapters of this glottology. We underline that all the steps are performed, and exist before the "historical" (here and now) event of antigenicity (e. g. the entry of a toxin): the matter of fact simply selects one of the possible channells, individuating the way: a kind of a generative process to form a phrase through a grammar. So the production of an anti-epitopic structure taken from the repertoire can be seen as the formation of a phrase using simpler units of sense.

Also the complex individuation of an imput from the external world performed by the nervous system is a definite channelling, each step of which consists in a relationship between molecular partners (the action of ionic pumps that allows the formation of the electric potential, the synaptic relation between neurotransmitters and their receptors, ando so on): the only relevant difference is that in the case of the nervous system the channelling is prefixed also in spatial terms as an anatomically defined network, while in the immune response the channelling must be determined each time through statistical encounters of cells.

6) - Semiotics and syntax: the syngeneic restriction

The repertoire contains also the anti-self sequences, and the organism is potentially capable to destroy itself through its immune system.

This event does not take place, because syntax includes rules that forbid tha channelling in the direction of what is self, and excludes the self as target. Interpreted either as deletion of clones or as a product of cooperative reactions, tolerance is obtained and mantained through a chain of molecular specificities: also the absence of reaction is a reaction.

Usually the specific trait of the immune system is considered the response against not-self. But two quite different aspects must be considered together: the response is both against not-self (therefore is a general exclusion of self: it is not a choice but a giving up of a large set of choices) and specific (against a given epitope).

In the case of antibody production, at the end of the channel we obtain a free anti-epitopic molecule, which can react with the correspondent epitope.

In the case of a cell-cell reaction (cytotoxicity) and in all the cell-cell reaction, an identification of "selfness" is necessary.

Each cell of the system, to cooperate with another cell, must be sure that it is matter of a cell of the same system (of the same organism or of an organism of the same individual characteristics) and deciphers on it some signs of "selfness" that mean "compatibility". This deciphering opens the gate, and the chain can begin. The same happens for the destruction of the cell-target: "foreing" is always interpreted in the frame of the "familiar", as if it were necessary to preliminary ascertain the use of the same grammar. In fact a language presupposes that the speakers agree to use that language, adopting its syntax and its semantic repertoire: the unknown which a language must always face (forming continually new phrases) is liable to interpretation because it is introduced in the frame of the known (which is old and well established, and is precisely the system, the mechanism of interpretation). The signals of selfness are genetically determined: in fact they must be invariant, not dependant on the "personal history" of the individual, or on environmental influences. They are constituted by MHC-coded antigens. The problem of presentation of an antigen toghether with MHC products (as two separate points, or as one point) is well known.

On the contrary, tolerance is an "historical" event, and depends on what the system finds during its development: it is not genetic but ontogenetic. Important to remark, also tolerance is restricted. This "view of differences in the frame of equality" is compatible with a concept of the immune system as phylogenetically developed mainly to mantain the integrity of the self against the tendency to disorder represented by mutations: the function of the system is to preserve the neatness of the genetic message from the ground noise, and the reaction against exogenous materials (bacteria, toxins, virus) would be only a side effect. Cells presenting "deviation" from the permitted scheme (self cells which become non-self), are eliminated, and for this is necessary that the exploring immune cell can before recognize selfness: it explores continually normal cells and individuates the altered ones confronting these with its (normal) scheme. So an immune cell is the touchstone of the self.

7) - Idiotypic network

The amino acid sequence of the antibody combining site is necessarily "new", therefore antigenic, and produces an immune response. The variable region of the correspondent antibody is the equivalent (in immunological terms) of the antigenic epitope: both (epitope and anti-antibody) can bind to the same amino acid sequence, which constitutes the variable region of the antibody. In its turn, the variable region of an anti-antibody is an epitope, and so on. In the whole immune system an "internal image" of the epitopic universe is present not only in the sense that the anti-epitopic sequences of DNA are present in the repertoire, but also that the system is provided with a universe of internal epitope-equivalents: the outside world has, so speaking, its representatives in the inside domain of the body. The production of anti-images is a continuous function of the system, which is always in a state of internal activity, a kind of at rest work. Therefore the input of an antigen can be seen also as a perturbation induced in a condition of equilibrium, forcing the system to reach another condition of equilibrium.

The character of an a priori mechanism is even more clear. But this does not mean that the immune system (or, generally speaking, an organism) is a kind of "autopoietic" monad, and that the external world is only a shadow: the system is derived from the outside and reacts towards the outside interpreting it. On the other hand, always we see that, in the course of phylogeny, a more precise and efficient interpretation of reality parallels a higher (and not a lower) degree of autonomy of the reader versus the reality to be read.

8) - What's semiotics for immunology (and vice versa)?

Immunologists can probably be stimulated by the concepts of semiotics (as by those of theories of information and of communication).

They are particularly sensitive to all the phenomena of signalling. Nevertheless the danger is high that they are tempted by a metaphoric use of the terms, that can confuse the argument. In fact semiotics is normally intended as a theory of human and cultural signs and codes. It involves symbolic activity, hypothetical structures, a formal logical frame and a certain degree of consciousness. Then questions of the following kind (perfectly meaningless) can be posed, and in fact are: if the immune system can make choices, or can build hypotheses, or has an intentionality in synthesizing its antistructures, and so on.

The different modalities in deciphering objective referents develop during natural history in absolute continuity, one from another, but they are distinct, and must be considered distinct. The continuity does not exclude sharp qualitative differences, and the existence of real "crises" in this pathway. A point of crisis is certainly represented by human knowledge, which is in continuity with, but cannot be reduced to, previous systems of meaning. A domain of modalities is also a domain of terms that we use about them, so it is necessary to define the meaning of the words and the context in which they are placed. In this perspective we have specified the term "sign" as a diadic entity which becomes triadic when a sufficient complexity is reached.

A symbolic conventional sign cannot represent the situation of the immune sign, or, even more so, of the "categorical" biological sign (which we have exemplified in the function of an enzyme molecule).

In any case, whether the concepts of semiotics can help the immunologist in his experimental work is, in my opinion, doubtful.

On the contrary, I think that immunology can give fruitfully its concepts to semiotics, to build a general theory of signs and codes, whose scheme we have traced starting from the origin up to the stage of human knowledge.

In our view, it is not symbolic language which "explains" the machinery of the cell, but it is this machinery that must explain (through a situable complexity) the nature of a symbolic language.

There is disagreement between those that consider semiotics as a pure human domain, and the present view of a general semiotic domain: the only question is where to place the threshold for sign. To avoid confusion, the term "protosign" could be used instead, up until the human sign, but this seems to me rather artefactual. In any case, the common points are that reality is interpreted by a priori systems provided by their own codes, and that all are a posteriori products of their natural history (and therefore of the reality that they can read).

BIBLIOGRAPHY

- G. Prodi: Le basi materiali della significazione. Bompiani - Milano 1987
- G. Prodi: La storia naturale della logica. Bompiani - Milano 1983
- G. Prodi: Biology as natural semiotics. IIId int. congress of semiotics - Palermo 1985
- G. Prodi: "Phylogeny of codes" and "Ontogeny of codes" in "Dictionary of semiotics" Sebeok ed. 1986

SEMIOSIS FOR THE IMMUNE SYSTEM BUT NOT THE IMMUNE RESPONSE OR WHAT CAN BE LEARNED ABOUT LANGUAGE BY STUDYING THE IMMUNE SYSTEM?

Edward S. Golub
Department of Biological Sciences
Purdue University
West Lafayette, IN 47906

Immunology and the Nervous System

Immunology, like grand opera, has always drawn on other disciplines. The elucidation of the structure and function of the antibody molecule, for example needed not only serology (the High Baroque foundation of immunology) but physical chemistry (the Classical period) and molecular biology (the avant garde). Similar uses of methods and advances in cell biology can be cited in the path toward the solution of the various problems of cellular immunology. Before the basic paradigms of the immune system were laid down there was much speculation about the similarities between the immune and the nervous systems. Because both systems are used to recognize and react to the outside world it was tempting to discuss the possibility that the two used similar mechanisms to solve the similar problems. I have discussed the shortcomings of this seemingly attractive notion elsewhere (1,2) and have argued that we now know enough about both of the nervous and immune systems to see that at most levels of analysis the connection is probably illusory. Granted, the immune system may be under some kind of influence of the nervous system, but it would be hard to name a system of the body that is not (see various papers in ref. 3). In this essay I address the central question of this conference; is the immune system, like language, a sign system?

Criteria for Semiotic Control

As Constance Jordan points out in her essay in this volume (4), the conceptual basis for understanding meaning in language was the analysis by Saussure of the word as a sign. A sign, which consists of the signifier and the signified, conveys meaning by standing for or indicating something else -its referent. Ordinary stimulus response systems are, by these definitions, not semiotic systems. However, a system composed of signs that in themselves signify entities other than what they themselves are, is semiotic. A forest ranger knows that 'where there is smoke there is fire' and does not wait until the flames are engulfing his tower to report a forest

NATO ASI Series, Vol. H23
The Semiotics of Cellular Communication in the Immune System
Edited by E.E. Sercarz et al.

fire. He sounds the alarm at the first sighting of smoke. Sailors knew of landfalls by spotting shore birds a day before sighting land (hence the well known cry, Bird Ho!).

Using these definitions and concepts, we must ask if the immune response is a semiotic system, ie, does it operate through the use of the sign. I will argue that when viewed from the vantage point of the <u>immune response</u> there is only stimulus response and therefore no semiosis. However, when viewed from the vantage point of the <u>immune system</u> there is a semiosis. Specifically, the immune response is stimulus response but the immune system is a network of interactions which uses signs.

<u>The Immune Response and the Immune System</u>

Immunologists have tended to use the terms immune response and immune system interchangeably, assuming that the immune response is the process carried out by the immune system. The usual view of the immune response/system is that of a set of cells, each with antigenic specificity expressed as a surface receptor which gives it the ability to react specifically with antigen. The specific interaction with antigen initiates a complex of cell interactions, mediated through interleukins, resulting in expansion of clones of specific cells and the expression of effector function (5). These cellular interactions must be regulated and the traditional view holds that the regulation is carried out by some as yet undefined balance between suppressor cells (6) and the immune network (7).

At this NATO conference, Constantine Bona made the important suggestion that it might be fruitful to separate the system from the response in our analyses. We think that this suggestion is of the utmost importance and propose to pursue it in asking if the immune response/system is under semiotic control.

We will define the <u>immune response</u> as the series of events which follows the interaction of the receptor on the surface of a lymphocyte with specific, exogenously introduced antigen. The introduction of exogenous antigen into the system results in the clonal expansion of specific helper and then effector cells resulting in immune function (antibody formation or cytotoxicity for example). There can be little doubt that this response is the result of the stimulus and that the stimulus is the antigen. These interactions which are admittedly complicated in biological terms, are a simple stimulus/response in systems analysis terms. Since stimulus/

responses are by definition excluded from semiotic analysis we must say that the response to antigen is also excluded.

But to say this does not mean that we can now all go home because there is nothing here for us. There is in fact a very interesting way of looking at the immune system/response from the perspective of the network (6). Although the network has been examined very extensively as a regulatory mechanism in the immune response, one of its most interesting aspects is its role in the maintenance of the steady state population of lymphocytes. Since each antigen combining site (paratope) is generated by random gene reorganization and has a unique structure, this structure forms a unique antigenic determinant (the idiotope). The idiotope can then act as an epitope to be recognized by the paratope of another antibody. Thus an antigen combining site has the potential to act as both antibody (paratope) and antigen (idiotope). The essential point is that there is a paratope in the system which recognizes an idiotope in the system. Jerne has suggested that the epitopes of anti-idiotopic combining sites generate internal images images which mirror the repertoire of epitopes of exogenous antigens in nature. It is these internal epitope/paratope/idiotope interactions in the absence of exogenous antigen which maintain the level of specific lymphocytes by constant interactions.

Thus, the internal image is a crucial feature of the network. This is an almost mystical concept and has caused problems among immunologists (see reference 8) and Jerne himself has helped add a note of mysticism by references to mirrors and what they can teach us about our inner self (9). But despite these problems the network and the internal image have become a normal part of immunological thought (and indeed, of our generating antibodies to receptors and other difficult epitopes). Since the internal image represents all of the antigenic structures to which the organism can respond reflected in the antigen combining sites of the receptors of the system, one can conclude that there exist in the internal universe of receptors on lymphocytes structures which signify some antigen from the external universe of antigens to other receptors.

Thus there is an <u>immune system</u> whose specificity elements are generated randomly (ie with no influence of antigen) and whose members communicate with each other in the absence of exogenously introduced antigen. When antigen is introduced into this system it initiates the <u>immune response</u> by combining with an antigen combining site on a receptor (paratope) causing the clonal expansion of specific cells. Since the antigen is reacting with a molecule which heretofore had been reacting with an internal image of the antigen (the idiotope) it follows that the same receptor recognizes and responds to both the antigen and the "sign" of the antigen, the internal image. From this it can be seen that the network is its own sign system which "sees" in itself the sign of the antigen.

What Do the Signs Tell Us?

Having said that the immune system in its base line state (the network) sees itself as the sign of antigen qualifies it as a semiotic or sign system. But now we must ask a very difficult question. Will knowing this aid in our understanding of the system? Will it allow us, for example, to formulate questions in a new way and allow us to approach the system more intelligently? To address these questions, we must first ask how valuable the idea of the sign as system has been to our understanding the function of language. What does not emerges from Professor Eco's book (10) or from Professor Jordan's analysis in this volume (4) is a notion that the study of signs (semiotics) has opened significant avenues toward understanding how we derive meaning from language. Problems of language such as contextual disambiguation (10) still remain. Because semiotics has not been the crucial key to understanding language why should we assume that it will be helpful in understanding the immune response? In fact, it might even be argued that the model for the study of the meaning of language which is more appropriate derives from a Chomskian (11) view of deep structure of language and the possible neuronal structures which are involved in language.

Indeed, the premise of this conference, the use of semiotics and language as a model for the immune system, may have put the question from the wrong direction. It may be that since the immune system is indeed a semiotic system, those who study language may now have the simple but very powerful model to put to constructive use. It may well prove to be difficult to think of a better model for the study of language than the immune system because it offers well-established paradigms and quanti-fiable tools. What we need now is the formulation of semiotic questions which we can ask of the immune system and see how the answers can help us understand the more difficult problems of communication through language.

References

1. Golub, E.S. (1982) Connections between the nervous, hematopoietic and germ-cell systems. Nature 299;483.

2. Golub, E.S. (1986) An overview of the immune system or an immunologists view of the nervous system control of the immune system. In, Frederickson et al Neuroregulation of Autonomic, Endocrine and Immune Systems. Martinus Nijhoff, Boston pg. 323.

3. Frederickson, R.C. A., (1986) ed. "Neuroregulation of Autonomic, Endocrine and Immune Systems". Martinus Nijhoff, Boston.

4. Jordan, C. (1987). A note on some basic concepts of the sign. This volume.

5. Golub, E.S. (1987) Immunology; A Synthesis. Sinauer Assoc., Sunderland, MA.

6. Green, D.R., Flood, P.M. and R.K. Gershon. (1983) Immunoregulatory T-cell pathways. Ann Rev Immunol 1;439.

7. Jerne, N.K. (1985). The generative grammar of the immune system. EMBO Journal 4;847.

8. Langman, R.E. and M. Cohn. (1986) The 'complete' idiotype network is an absurd immune system. Immunol Today 7;100.

9. Jerne, N.K., (1984) Idiotype networks and other preconceived ideas. Immunol. Rev. 79,5.

10. Eco, U. (1984) Semiotics and the Philosophy of Language. Indiana Univ.

11. Chomsky, N. (1985). The Logical Structure of Linguistic Theory. Univ. Chicago Press.

ACKNOWLEDGEMENT

I thank Constance Jordan, Doug Green and Eli Sercarz for stimulating discussions. Although no experimental work is presented in this essay, much of the impetus for the thought comes from work which is supported by NSF DCB 8411716.

DOES THE HUMAN MIND USE A LOGIC OF SIGNS DEVELOPED BY LYMPHOCYTES 10^8 YEARS AGO?

F. Celada
Department of Immunology
University of Genova
Genoa, Italy

During the months preceding the first meeting on Immunosemiotics, I tried to work on one of the admittedly vague intuitions from which our endeavour originated, i.e., that some evolutionary link must exist between the general behavior of different systems of the same organism - such as the immune system and the brain - which have in common the genetic background and the capacity or the assignement to offer well-balanced responses to the signs and stimuli of the environment.

Conversations with our semioticist counterparts and the reading of a few essays written by them certainly did not make things easier for me, as I was confronted with a body of definitions and rationalizations of the language of signs much too large to be readily applied to biology. This difficulty convinced me that the initial hypothesis had to be tested not by trying to apply human interpretative standards to the lymphoidal system but rather by comparing the two modes of response when confronted with a "critical" situation.

The same conversations and readings revealed a predictable difference between the reasoning of semiologists and of us biologists, namely we are much more impregnated with Darwinism. To us it comes as a matter of course to relate observed behavior in man to that of species lower in evolution (and also of cells and organs which constitute the organism under test), and it comes naturally, in the presence of a complex response, to try to identify a selective pressure important enough to have motivated its coming about. Underlining the need for a selective agent privileging the most efficient and useful responses could be a modest contribution of the biological moiety to the present interdisciplinary meeting. In simple organisms like bacteria the environment selects the fittest out of a population that is only apparently homogeneous, but in reality quite "diverse". Thanks to their flagella bacteria are endowed with motility. Most will move at the same speed but some (e.g. by mutation) have acquired the capacity of moving faster, others of moving more slower. The behavioral property "fast movement" could turn out to have a selective value for the survival of the bacterium in certain environmental circumstances. As a result, in the next generations the progeny of the few

NATO ASI Series, Vol. H23
The Semiotics of Cellular Communication in the Immune System
Edited by E.E. Sercarz et al.

fast cells will overgrow the slow ones. A superficial observer SO may mistake this change for a learning process at the level of the bacterial colony. Now let us suppose that instead of strictly unicellular organisms, the objects under SO's microscope were cells either constituting a multicellular organism MO, or constituting organs or systems within MO: SO may conclude that MO has learned to "move fast", and he may even be right, because it seems theoretically feasable that selective processes at the cellular level may result in a "learned" behaviour of the organism. Examples - each of them characterized by a powerful selective environmental force - are a), the addiction to drugs and poisons (back to the case of Mithridates) or, b), the resistance to the immunosuppressive drug Cyclosporin. The latter case is particularly well aimed: it has been recently demonstrated in my own lab that graft-rejecting T cell clones have a widely different sensitivity to the immunosuppressant (1). The selection operated by Cyclosporin will privilege the most resistant clones, causing the entire immune machinery to become less sensitive to the drug. It is interesting to note that a competent evolution for the betterment of the immune system may in the end yield a very unhappy result: the rejection of a kidney graft, that may even cause the death of the organism. But the IS is innocent: how could it imagine that alien-looking, green clad surgeons would be so crazy to introduce a foreign body on purpose?

A further step is to ask whether behind human learning there are Darwinian processes whereby under pressure - and what kind of pressure in this case? - certain cells and certain synaptic connections are favored over others. Fascinating as it may be this seems to be rather uncertain ground, since neurons do not multiply in the mature brain; selection without proliferation is hard to imagine and to test, while the embryo's nervous system may be difficult to study for other reasons.

The place where Darwinian selective processes occur and can be studied in great detail with modern tools is obviously the Immune System. Is it meaningful to take it (the Immune System) as the testing ground of the propositions stated in the beginning?

Against this idea stands the conservative objection that both the cells and the IS as a whole lack conscience. This objection has been fortunately dispelled by Umberto Eco in his opening statements: conscience is not a prerequisite for using the language of signs.

Another serious objection is the difference of cell lineage. Admittedly the thinking brain and the neurons are not on the same maturation-differentiation track as the lymphoid cells. However, since they share the same genetic backbone which originated in the zygote, the hypothesis - or should we say the abduction - that certain ways of behavior may have been selected in parallel tracks could be held. Although I realize that I shall not be able to prove it, at least I like to bring some more water to this mill.

A typical "extreme case" for the reasoning man is his response to ambiguity. We have learnt that ambiguous signs are everyday events, and rather humorous and puzzling examples have been cited during this meeting, using that very modern and sophisticated sign system that is language. It is in fact impossible at face value to decide which is the right interpretation of sentences like the one proposed by Eco (2) in Italian "La vecchia porta la sbarra" or "John makes love to his wife twice a week; so does Charles".

I shall briefly repeat the semioticist's analysis of the resolution of the ambiguity of the last sentence. The receiver of the message needs some additional information before he can favor one or the other of the possible interpretations. Since the message itself cannot be modified the additional information must concern the sender of the message (the author of the sentence), and the receiver's reasoning will take the form of two alternative abductions: if the sender is a notorius gossip then the interpretation will be that John's wife did commit adultery; if instead the sender turns out to be a serious statistician studying the sexual behaviour in a certain population, the interpretation will be that John and Charles love each their own wife.

I shall add - as the self-appointed official evolutionist - the concept of motivation/selective pressure. If the receiver is careless and does not properly investigate his sources, he faces the danger of the wrong interpretation. This may well be a matter of small account most of the times. But it is easy to imagine circumstances where making the wrong decision could be serious. If John and Charles are members of the Court or of the Royal family during the Renaissance, the mistake may cost the receiver his head: in fact, a similar misinterpretation of complex signs prompted Otello to strangle Desdemona and kill himself. In our days it could cause the loss of a job for a journalist and ruin his career. If John and Charles were in the financial world even the stock exchange would react with amplified consequences. Hence the fear of mistake. Fear is a great motivation and certainly causes people to try to reach the right interpretation at any price: a typical selective situation where only the fittest will survive and step forward in society.

Let us now make the bold transition to the immune system (IS) and begin our analysis at the level of the single B lymphocyte. This cell is the typical representative of the evolved IS. It carries on the surface a number of molecular tools of immunoglobulin nature - receptors - which can interact with other molecules of the environment in a specific way: i.e., the receptors characterizing the progeny of a single precursor cell (a clone) differ from those of all other clones and are able to react biochemically with a small proportion (to the limit: one) of the universe of environmental molecules that are called antigens. The interaction antigen-receptor is necessary for the elicitation of a response by the B lymphocyte. The response will consist in a series of changes within the cell, conducive to its proliferation (one cell will give rise to several hundreds) and differentiation, culminating in the massive production and secretion of antibodies. But take notice: this response is not automatic: after having bound antigen the B cell can also produce another response: no change - no proliferation - no differentiation. This possibility of choice, however limited, shows some similarities with the interpretation problem discussed above in man: also here there is ambiguity in the sign that is received. The epitope, the molecular region making contact with the receptor could belong to a foreign antigen, signalling e.g. the invasion of dangerous microorganisms; in this case the fittest response of the B cell should be yes. On the other hand it could as well be part of an autoantigen, a molecule derived from the same organism: in this case the proper response should be no, to avoid dangerous autoaggression. A mistake in this decision is certainly serious. It may result in the death of the cell, of the individual, even of the species. Autoimmunity has been even proposed as a possible cause of the extinction of dinosaurs. Confronted with this formidable selective pressure the immune system has evolved an efficient way of decision making. The B cell must gather independent information about the real nature of the antigen. For this purpose the IS has produced over the last 10^8 years a second type of cell, the T helper. This one - also in a specific way - recognizes a different epitope of the same antigen molecule, provided it is presented on the surface of a cell of the same organism. The presenting cell is classically a macrophage, but it could also be the B cell itself. The testimony of the T cell is also given in the form of a yes or no response and is reliable: when the Th responds it means that the antigen is foreign and potentially dangerous. This is because the T cells have been properly educated in the thymus where the autoreactive T have been eliminated or silenced. The signal of the activated T cell is for the B lymphocyte the proper "additional information": if it arrives, the B will put an end to hesitations, and begin its action.

Figure 1

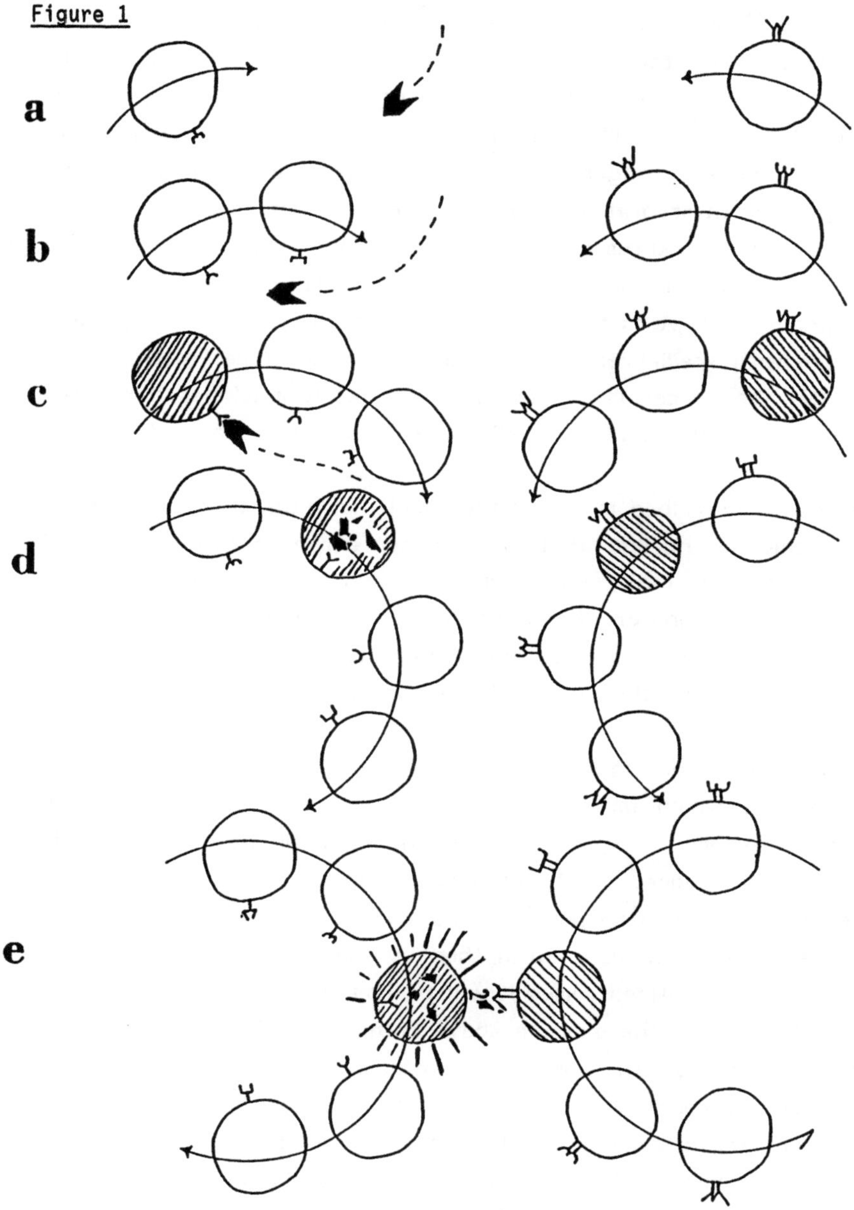

LEGEND TO FIGURE 1
The mill of the B cells (left) and the mill of the T cells (right):
a) antigen appears
b) antigen approaches
c) antigen is recognized by one of the many B cells
d) internalization: but what action should be taken?
e) decision is made when one of the many T cells certifies that the antigen is ok to react to.

Table 1 and Fig. 1 show an example of T-B cooperation of the type described above. The data are from the work of Lanzavecchia (3). Both the B and the T cells are clones, therefore exhibiting a single specificity towards the antigen tetanus toxoid (TT). The experiment shows that in the presence of an anti TT B cell and the antigen the anti TT Th cells proliferates (Figures in Table 1 represent incorporation of radioactive thymidine as a measure of the proliferative activity, expressed in counts per minute, (CPM). A condition for the T stimulation is that the B lymphocyte that binds and concentrates TT on its immunoglobulin receptors be allowed to internalize, process and present TT in connection with specialized molecules on its surface (the "class II MHC molecules"). After the Th cell is activated, it secretes lymphokines that reach its collaborating B lymphocyte and stimulate it to produce and secrete antibody.

The T-B cooperation described is one of the functional units that make up the immune system: the total number of these components may be between 10^8 and 10^9. It corresponds to a single neuron with its dendritic connections or to a single transistor in a computer. Being so tiny it is understandable that its highly sophisticated decision-making mechanism is however limited to a yes-no, 0-1 response and unable to choose between qualitatively distinct alternatives.

But we can look one level higher. Not at the entire IS, but to a limited population of B cells which has been preimmunized with the native antigen GZ, - a bacterial enzyme, /B-galatosidase - and contains now memory cells belonging to different clones, practically directed to all "native" epitopes of GZ. It so happens that among all possible antibodies - each the marker of a B clone - we are able to distinguish two of them by their functional properties, anti-G and anti-Z. When antigen comes it will be bound by anti-G, anti-Z and anti-n B clones but, as in the example above nothing will happen before the puzzle of ambiguity is solved. Now let us introduce into this system one particular Th clone, recognizing a particular part of GZ. We have recently found (3) that contrary to expectation this clone will not offer its cooperation to all B; moreover, when we look at the final ab specificities, there are T clones that will favor anti-G over anti-Z, and other clones that will do the opposite. The introduction of either the first or the second kind of Th (let us name then i and ii respectively) will trigger the B cell population to produce either anti-G or anti-Z: thus, within an already selected population we can see the IS's choice between two different yesses instead of between yes and no (Table 2 and Fig. 2).

TABLE 1

Yes/No choice. The B clone expands and produces Ab when the Th cell provides a second signal concerning the antigen

Clonal B cells anti TT*	TT antigen (ng)	Clonal Th cells anti TT	Thymidine (CPM/culture)	incorporation Produced Ab
10^4	0	0	0	-
10^4	10	0	0	-
10^4 x irradiated	10	10^3	8000	-
10^4	10	10^3	12000	+
10^4 fixed	10	10^3	0	-
10^4	10	10^3 x irradiated	0	-
0	10	10^3	0	-

* TT = Tetanus Toxoid

TABLE 2

Alteration choices. A B cell population consisting of several anti GZ clones produces antibodies of one or the other fine specificity according to the Th clone introduced into the culture

B anti-G + B anti-Z + B anti n	Ag (GZ)*,ng	Ti	Tii	Tiii	Ab produced anti-G	anti-Z
10^5	10	0	0	0	0	0
10^5	10	10^3	0	0	++	+
10^5	10	0	10^3	0	0	++
10^5	10	0	0	10^3	0	0
10^5	10	10^3	10^3	0	++	++
10^5	10	0	10^3	10^3	0	++
10^5	10	10^3	10^3	10^3	++	++

* GZ is E.Coli ,B-galactosidase used as antigen.

G, Z, n represent epitopes on the GZ molecule.

Ti, Tii, Tiii represent different Th clones, capable of providing help to anti-B, anti-Z, anti-n B clones.

Figure 2

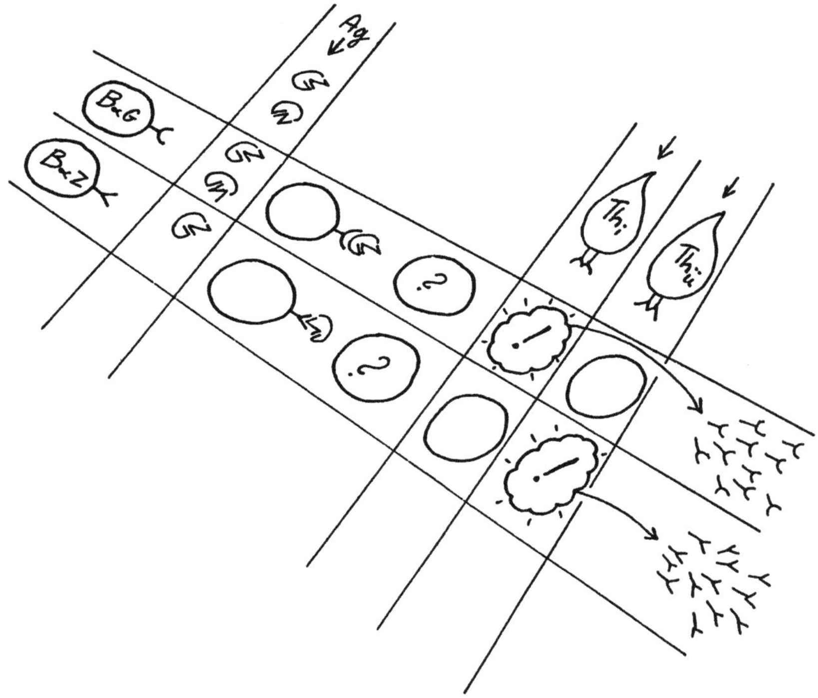

LEGEND TO FIGURE 2
On the track of a population of antigen specific (but with a difference in fine specificity B cells, a number of encounters.
First, antigen arrives: GZ. As a consequence GZ is bound according to each B cell specificity. After, a time of hesitation (question marks): the Th arrive: if Thi, then B anti-G is stimulated to go to the response routine. If Thii, then B anti-Z is stimulated.

It seems to me that this experiment has some deeper additional implications. A plausible origin of the preferential connection between i and anti-G and ii and anti-Z lies in the processing-presentation step. We have already seen that in memory-triggering, unlike during the primary response, it is mainly the B cells that - having concentrated antigen (GZ) on their Ig receptors - internalize it, reexpose it in conjunction with Ia antigens and then present it to the Th. It is our hypothesis that anti-G and anti-Z when they take GZ into their respective B cells somehow influence the processing of different processed fragments which are then displayed on the surface. Thus Thi may recognize what is shown on anti-G, Thii may recognize what is shown on the surface of anti-Z. The stimulation provided by the two Th clones may result in directional cooperation (see Fig.2). If this mechanism were general, then the decisional role of the B lymphocyte would be considerably enhanced. A single cell or clone constituting 10^{-8} of the B universe in the IS would - as said before - recognize the antigenic sign, then pause in hesitation needing to wait for a second reassuring piece of information about the sender. But - and here is the added step - this pause is not passive: the cell seems to feel the urge of getting its second sign. It works on Ag in a very "personal" way, in a way governed by the most private identity of the clone, the specificity of the Ig receptor. When it has done it, it goes around chasing, for the connecting Th, a Th seeking not any form of GZ, but that which is "processed in the presence of anti-G". It is of some interest that obtaining the reaffirmating second signal has required a transformation of the offered first sign. The second signal for the B cell's further activation. Our cell has gained information, thought about it, and used the processed product to allow development. This active decision making is - to me - the hint that something we - men - may have learnt from our little components, a method to survive through knowledge. If this were only partially true what we know about the IS may be applicable to the mind.

Literature references

1. Manca F., Carrozzi S., Kunkl A., Barocci S., Dessì V., Valente U., Fontana I. and Celada F. Transplantation, 41, 2, 119, 1986.
2. Eco U. personal conversation.
3. Lanzavecchia A., Nature, 314:537, 1985.
4. Manca F., Kunkl A., Fenoglio D., Fowler A., Sercarz E.E. and Celada F. Eur. J. Immunol., 15, 345, 1985.

Acknowledgements

I wish to thank Eli E. Sercarz for reading the manuscript and Luisa Di Rosa for the word-processing. Work in the laboratory discussed here was supported by C.N.R. grants n° 85.1792.95 (P.F. Chimica Fine e Secondaria) and n°86.01415.57 (P.F. Tecnologie Biomediche e Sanitarie) and by the Italian Ministry of Education (40% grants).

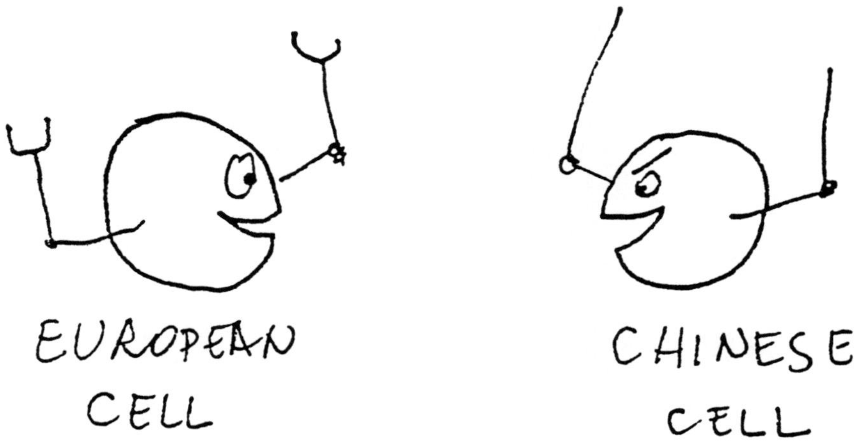
EUROPEAN
CELL
CHINESE
CELL

Contexts in the cellular communications: How the mobile cells recognize right partner cells to talk with?

— To Richard K. Gershon —

Tomio Tada

Department of Immunology, Faculty of Medicine, University of Tokyo
Tokyo, Japan

Every message is made of signs; correspondingly, the science of signs termed semiotics deals with those general principles which underlie the structure of all signs whatever, and with the character of their utilization within messages, as well as with the species of the various sign systems, and of the diverse messages using those different kinds of signs.

Roman Jacobson

NATO ASI Series, Vol. H23
The Semiotics of Cellular Communication in the Immune System
Edited by E. E. Sercarz et al.

Introduction

Having learned from Sir MacFalane Burnet, cellular immunologists see the immune response as a sociobiologic phenomenon where dynamics of interactions between constituent cells finally make the decision of the system itself. It is also analogous to the nervous system where an adequate response is made through the communication between the cellular components whose interconnectedness is precisely predetermined. Unlike the hepatic parenchymal cells, the cells in both systems are characterized by an enormous diversity and polymorphism, and exchange signs and signals to produce a meaningful response.

The immune system is of particular interest, as the different subpopulations of lymphoid cells constitute cellular circuits where signals are transmitted from one cell type to the other in an unmistakable and meaningful way even though these cells do not make physical synaptic contacts as in the nervous system. Antigenic input generally induces a well regulated train process where different sets of lymphoid cells are sequentially activated to produce an output that neutralizes the initial input (antibody production or suppression of the response).

The purpose of this paper is to redefine several aspects of cellular communications involved in the immune circuit under semiotic considerations, and to view this regulatory system under a general biologic rule dealing with the production and reception of signs. Immunosemiotics is thus defined as the study of the general principles underlying the structure of sign systems perceived by different cells within the immune machinery. My intension here is based on the considerations that all the terms used to describe cellular communications in immunology have never been precisely defined (sign, signal, recognition, communication, information, message, etc.), and that semiotics may provide important clues to correct and restate what we have been abused to use in describing immunocyte interactions.

1. Premises for the immune circuit

Through the analysis and definition of different functional roles of T cell subpopulations, an idea has emerged that all the functional sets of T cells are mutually connected to constitute regulatory circuits which are programmed during the early ontogeny of T cells (1-3). This concept is often referred to the "immune circuit" in an analogy to the nervous circuit. The circuit idea has a clear distinction from the network: The network primarily deals with the interaction between different idiotypes of immunoglobulin V domains, either recognizing or being recognized, and regardless on what cell types these idiotypes are expressed. The circuit is much more concerned with the interactions between functionally different cell types regardless what devises are utilized for mutual recognition. Best examples are seen in Gershon's schema (1), where antigen-reactive suppressor inducer T cell (Tsi) produces a factor which turns on the suppressor effector T cell (Tse). Tse then produces the effector suppressor factor (TseF) which suppresses both helper T cells (Th) and Tsi cells by which Tse has been induced. In our own experimental system, we have described the existence of another cell type (transducer T cell) which receives the inducer signal from Tsi and transduces Tse from a precursor pool (3). Another distinct cell type which prevents Th from the effect of Tse is designated as contrasuppressor T cell (Tcs) (4). The point in this schema is that all the member cells have distinct cell surface markers and perform only prescribed functions but not others.

There are several examples of preprogrammed multicellular interaction models explaining the antigen-driven suppression or augmentation of the immune response (1-6). Our own model of multistep regulation of the secondly antibody response was formulated by the observation of sequential activation of different cell types initiated by antigen, which finally produces an ultimate suppressive or augmenting signal for Th cells (3). Others also demonstrated a cascade activation of distinct cell types, which is linked by idiotypic determinants and producing amplified suppressive signals (5,6). These models have suggested the existence of general rules for the methodical activation of different sets of cells, which finally determine the class of effector responses either to be inductive or suppressive. Although these considerations of multiple member cells each having only a prescribed function are not well accepted nowadays, we must still admit that at least different functional subsets of T cells are exchanging antigen-driven signals according to the general rule of selecting their appropriate partner cells. Such methodical and predictable responses of different sets of cells could not be easily explained by the neutral idiotype network which has no such to blue print.

There are several theoretical premises for making the immune circuit physiologically functioning. The immune circuit is composed of different lymphoid cells, mobile, recirculating and continuously recycling. They belong to different subpopulations; helper, suppressor contrasuppressor, transducer, amplifying etc. They are different with respect to their prescribed functions and cell surface markers. Accordingly, the functional molecules they utilize for cellular interactions, and the receptors and acceptors for such molecules should be unique to the given cell type. From the pure theoretical considerations, all the member cells have to have only one functional role for each at least at a given time in their life span (functional singularity). Thus, the quality and content of the signal produced by the cell type is inherently determined being specified by its gene expression. The member cells in the circuit have to know exactly the right partner cells to interact with. The selection of these partner cells can be made independently from the antigen specificity, as one cell type is programmed to interact with another cell type regardless of the antigen to which these cells respond (program of cell interaction). This program should be made through the recognition of an isomorphic structure unique to the given cell type by the other, which finally determines the direction of message conveyance. Thus, the direction of cell interaction is always preempted by the genetically prescribed function of the subset, as one cell type can talk only to his partner but not in a opposite way at one time (unidirectionality). The activated cell may be able to talk back to the activator cell only after a proper activation step has preceded (feedback mechanism). Finally different classes of immunoregulatory events, suppression, augmentation, contrasuppression should utilize different sign to activate different circuit producing different output. These circuits should be interconnected to construct a greater integrated circuit, i.e., the immune system. The blue prints of circuits should be determined during the early differentiational history of lymphocytes by an adaptive process.

Some of the above premises seem apparently wrong, as we already know that some Th cell clones have other functional properties such as cytotoxic killing and multiple lymphokine production. I shall notwithstanding disregard them in describing the basic principles of selective cell interactions to convey signals within the circuit, because in this theoretical trial we see these cells only as abstruct subjects which handle different signs and signals. Accordingly, the following discussion will concern several facets of immune circuit only on the semiotic perspectives but not on detailed experimental fact.

2. Triadic relationship between antigen, macrophage and T cells

Antigen is often referred to as the initial sign to activate T cells, but the processing and display of antigen by macrophages or antigen-presenting cells (APC) is absolutely necessary for the recognition by T cells. Although the antibody molecule, that is an output of the system, can react with unprocessed antigen molecules in ridding them, the system is never turned on by the crude antigen itself. Antigen-presentation by APC is, therefore, an essential process of denotation of the sign, only after which T cells can perceive it. Now the antigen is signified to T cells.

This denotation of antigenic signal by APC is a complex event where fragments of antigen are associated with a self major histocompatibility complex (MHC) product which T cells are programmed to recognize (associative recognition). Jan Klein simply defined MHC as a group of genes coding for molecules that provide the context for the recognition of foreign antigens by T lymphocytes (7). The complex structure of the T cell receptor, as we see now, should correspond to the context and content of the sign given by the sender (APC). The T cell receptor recognizes the content of sign (antigenic epitopes to be recognized by T cells) in context of class I or class II MHC polymorphism, which determines the appropriate class of the effector response being properly regulated by other cell types. If the MHC is mismatched, T cells cannot recognize antigen as a sign as it is spoken in an unintelligible context. The unknown molecular events occurring at the antigen-presentation, referred to Ir and Is gene effects, lead to the consequence to either positive response or specific suppression of the response to the antigen. Thus, the process also contains a certain connotation added by the sender (APC) of the sign. These events altogether make up a message on the surface of APC which T cells perceive. Some messages are meaningful only to those having CD4 or CD8 molecules as supplementary devises for recognition, which also influence the subsequent quality of the response.

This triadic relationship between antigen, APC and T cells makes the initial recognition step as a subject of semiotic considerations, as it is clearly different from the direct cause-effect phenomenon such as the response of muscle fiber to an electric stimulus. The sign is perceived by T cells only if it is signified by APC where several steps of possessing and representation have preceded. In addition, the recognition of antigen by T cells seems to contain several other processes of semiotic interests; these include the decision which is concerned with the acceptance or refusal of the sign given by APC and/or T cells, interpretation of a complex and unintelligible sign into a perceivable sign by APC, transduction of a sign into nonspecific responses (cytokine production), choice of receiver of the sign (CD4 or CD8 positive cells), preparation for a counter signal, etc., all of which are subject to the parallel consideration by both semiotics and experiments. As the cognitive processes in the immune system are apparently redundant and prone to make mistakes, they can easily result in a formation of misleading information, even in a lie, which is also an interesting subject of semiotic examination.

3. Construction of the minimal regulatory circuit

On the basis of above consideration of APC - T cell interactions, let us examine other antigen-driven consequences in the immune system. T cells receiving signs on APC are activated, and now can act as Th cells. They recognize the partner B cells by looking at the identical sign which T cell have seen on APC. Here, again B cells display antigenic material in association with class II MHC product to signify it as a meaningful sign for T cells. Whether or not the denotation of antigen by B cells is the same to that of APC is not known. If this denotation of by B cells was successful, T cells are re-activated, and then they counter-activate B cells to produce an output signal, i.e., the immunoglobulin (antibody) synthesis. This makes a new triadic relationship between APC, Th and B cells. The epitopes seen by the immunoglobulin are not necessary to be the same as those recognized by T cells. In the context of network, an external antigenic epitope can be redefined in a semiotic term as a word spoken in a foreign language that can be translated into the sign (idiotope) which the immune network has been using in its dictionary before this word was spoken.

However, the activated T cell talks not only to B cells, but also to other T cells. This initiates the formulation of another triadic relationship between APC, Th and Ts cells, which constitutes a minimal regulatory circuit.

One of the simplest examples of circuit activated by a T cell clone (8) is as follows: We have established a self class II-restricted helper T cell clone derived from a mouse strain having $H\text{-}2^k$ MHC. It has been cloned and maintained by stimulation with $H\text{-}2^k$ APC plus antigen followed by cultivation with a T cell growth factor, IL-2. After a long term cultivation, however, the cells were found to have lost the antigen-specificity and became an auto-class II reactive Th cell clone. The clone was able to help B cells polyclonally to induce proliferation only when these B cells have the relevant class II antigen (A^k molecule) regardless of their antigen-specificity. The clone could be stimulated by self APC without any foreign antigens. Of particular interest was that the clone had a very strong stimulatory activity for T cells only when the latters were derived from strains sharing the same class II gene (A^k). The T cells proliferating upon stimulation by the clone was turned out to be Ts (see below). This gave us a unique opportunity to study how T cells can find the appropriate partner cells in the immune circuit regardless of the fact that most murine T cells express no class II antigens. As the observed activation of T cells by the cloned Th and the subsequent regulatory activity of newly induced Ts are all strictly H-2-restricted, they should recognize isomorphic structures other than class II antigen marked on corresponding functional T cell subsets.

Observations are as follows: We found that the injection of the above self class II-reactive T cell clone into syngeneic mice resulted in the production of a potent Ts in the spleen of the recipient. Ts thus induced was able to suppress normal Th of the same strain but not of H-2 incompatible strains. There was a strong MHC restriction both in the induction and effector function of Ts. No antigen-specificity was observed in both afferent and efferent phases: For example, Ts induced in $H\text{-}2^{kxb}F_1$ could inhibit the Th from $H\text{-}2^k$ but not from $H\text{-}2^b$ regardless of the antigen-specificity. As murine T cells of both helper and suppressor classes do not express conventional class II antigens on their cell surface, the question is asked how the A^k-reactive T cell clone could activate the Ts that could inhibit only the A^k-restricted Th regardless of the absence of A^k molecule between these two cell types.

The experiments using both Ts and Th cells obtained from various radiation bone marrow chimeras finally gave the answer: Th cells derived from chimeric animals produced by the transplantation of F_1 bone marrow cells into lethally irradiated $H\text{-}2^k$ parent (designated as $F_1 \rightarrow H\text{-}2^k$) are known to possess only the $H\text{-}2^k$ restriction specificity but not the others. This is due to the adaptive learning of the $H\text{-}2^k$ restriction specificity during the ontogeny of T cells in the thymus where thymic epithelial cells express only $H\text{-}2^k$ but not $H\text{-}2^b$. $F_1 \rightarrow H\text{-}2^b$ Th is accordingly $H\text{-}2^b$-restricted. Both Th can equally help F_1 B cells to mount a secondary antibody response, as class II antigens of $H\text{-}2^k$ and $H\text{-}2^b$ are codominantly expressed on F_1 B cells. If we take Ts induced in F_1 by the injection of the above $H\text{-}2^k$-restricted T cell clone, it could suppress the response induced by $F_1 \rightarrow H\text{-}2^k$ Th but not that induced by $F_1 \rightarrow H\text{-}2^b$ Th. Since all the member cells in this experimental condition are genotypically of F_1 origin, it was concluded that Ts recognizes the restriction site on Th from the chimera, which was adaptively acquired during the time when Th cells developed from F_1 stem cells in the thymic environment of $H\text{-}2^k$.

The simple deduction is that some T cells, especially Th cells, are taught to read the context of signification on APC (antigen presentation with class II MHC), but other T cells are programmed to recognize different signs which are generated in Th cells by the recognition of the MHC context. Ts are thus programmed to recognize these second signs which are dictated by the context of first sign. These newly generated signs should correspond to the <u>interpretant</u> of Peirce, and can theoretically produce further signs infinitely. The interpretant is dictated by the initial context, and therefore can be used to convey the message to the next receiver with certain connotation added by the interpretion process. Thus, whenever antigen is introduced, the first set of T cells receives the sign (antigen in context of self MHC) on

APC, and generally produces a new sign (interpretant) which can be perceived by the second set of T cells. Although the nature of this newly produced sign by Th is not precisely known, it is undoubtedly related to the structure that recognizes the self MHC, as the context of recognition by the first set is being recognized by the second set. However, it is independent from the antigen-specific structure (conventional idiotype of T cell receptor), as the activation of Ts is antigen nonspecific. Such antigen-nonspecific but cell type-specific structures should be isomorphic rather than polymorphic. It is conceivable that Ts perceives the interpretant from Th which was produced by the recognition of context (MHC restriction specificity) rather than that of the content (idiotope or paratope specified by antigen). This makes the model different from the conventional network, as the network is much more concerned with the content of signs rather than the context.

We have recently learned that the first set of T cells (MHC-restricted Th type) induces not only Ts but also its counterpart, the augmenting T cell (Ta), which amplifies the antibody response in an antigen-nonspecific but H-2-restricted pathway (9). The presence of this cell type allows the parallel activation of another regulatory circuit that augments the immune response. Ta is also activated by the MHC-restricted recognition site of the Th cell type. The reason why Ta has not been documented previously is that the coexisting Ts is in many cases predominant over Ta under the physiologic condition.

All three components, Th, Ts and Ta are similarly MHC-restricted, but the meaning of restriction is different in Th from Ts and Ta. As pointed out by Dr. Mitchison (in this volume), the restriction of Ts and Ta are secondary to the interpretant of Th, and can be better designated as "epirestriction"; MHC restriction of Th type dictates the epirestriction of Ts and Ta in the early differentiative history in the thymus, which programs the suppressor and augmenting circuits.

What we see in the present model is amazingly simple: In the antibody response for any antigens, the immune system can utilize the built-in regulatory circuits in which MHC-restricted Th activates Ts or Ta that is MHC-epirestricted. Whenever Th is turned on by antigen, it subsequently activate these regulatory T cells that in turn suppress or augment the Th to modify the output of the system (Fig. 1). Some antigen may activate the Ts circuit rather than Ta resulting in an unresponsiveness, while the others induce an accelerated response by the aid of Ta circuit. Under the physiologic condition, both pathways may be sequentially or concurrently activated to produce the normal self-limited or emphasized response. The signal to determine the switching of this flip-flop machinery should be the key to understand the tolerance vs induction.

Construction of a concept

We can still go on and on to redefine several immunological phenomena in semiotic terms, but I shall stick on to the above simplest model to construct a minimal concept of circuit by an epistemological way. As we have discussed, the circuit idea has clear distinction from the network. It does not concern the issue of repertoire generation, but asks how the mobile immune cells of different classes can make unidirectional and meaningful communications to produce an output of a defined quality. As the network does not discriminate the recognizing and recognized, there is no way for the system to determine the direction of the response. There are always two open ends which can be either the entrance or exit. Any of the circuit model one may choose is concerned with the unidirectionality of the cellular communications which can only be achieved by the signification of a sign from one cell type to the other. In the simplest model, the antigen is signified to the second, and second cell interpret the first sign to talk to the third. The third can either produce a next output or talk back to the second to shut off the sender of the signal. If we examine many of more complex models, we always see the repeats of the same communication pattern of three individual cells (triadic domain). The antigenic input can activate the first triadic domain (upper dotted triangle in Fig 2). The signal could be conveyed to activate the second domain, but can also be diminished from the whole system by neutralizing the initial input. By the sequential activation of domains, the signal finally produces the output, i.e., specific antibody formation or neutralization (suppression) of the signal. Thus, in the circuit model there is only one entrance and one exit.

There exists a so called "nest box" structure where the system is composed of many smaller repeats of above basic three cell types each making triadic domains by interacting with connecting cells (Fig. 2). The principle of interaction is always the same; T cells have a capacity to recognize the context of sign, which makes them to perform selective communication with the other. The context is always translated to the interpretant that is recognized by the next, which determines the direction. This can easily be analysed by a mathematical model such as autoregressive multivariate model, as the basic principle is a built-in automatic control with a feedback.

Ample evidence now indicates that I-J and Iat, which have been mistaken as the products of MHC, are examples of the interpretants of class II antigen (10-12), and that they are associated with the class II-recognition site of Th type cells. They are essential for the activation of the second order regulatory T cells. Indeed, if we remove I-J positive T cells among antigen-primed splenic cells, both MHC-restricted Ts and Ta cannot be generated against any antigens (9). Since it has been demonstrated that both I-J and Iat epitopes are adaptively generated in radiation bone marrow chimeras during the differentiation of T cells under the different MHC environment (10 -12), the domains of the MHC-restricted minimal regulatory circuit are programmed at the critical time of T cell ontogeny. Thus, the immune circuit is predetermined just like the neuron-by-neuron circuit in the nervous system before it encounters with antigen. The important property is that the program is not embodied within the genetic library, but is somatically formulated according to the environmental MHC with which T cells develop their recognition abilities in the thymus.

Our view of this programmatic process of immune circuit is as follows: The development of the immune system depends on the generation of a certain set of antecedent cells which are initially selected to recognize self MHC. A large portion of programmatic phenomena follows in a manner that other consequent cells are obliged to recognize the antecedent cells. Such a program should be made through the interaction of isomorphic structures somatically generated in the antecedents and consequents rather than of idiotypic structures, as the latters are different from one clone to the other making the program unpredictable.

We have proposed a model to explain the above situation as illustrated in Fig. 3: Thymic educator cells express MHC antigen as pleiomorphic structures which are later used as the context to present antigenic signals of T cells. The epitope on this prototypic molecule used for education is called "prototope" which select Th cells. The epitope on Th cells specifically involved in the recognition of prototope is designated as "antetope", as this cell type is the antecedent for the development of the regulatory T cells. Those selected for the recognition of "antetope" are Ts and Ta that carry "metatopes" complementary to antetopes. These minimal members should suffice the formulation of the minimal regulatory circuit, where antetopic T cells recognize antigen in context of prototope and activate metatopic regulatory T cells. Both suppressor and augmenting circuits are explained by these minimal member cells. More complex model may also be formulated by the extention of this minimal model, as the triadic domains can be multiplied and integrated into the whole system.

These MHC-restricted circuits are apparently operative independently from the conventional network because we see no antigen specificity in the construction of circuits. Basic assumption is that Th cells activated by any antigen can equally be regulated by these built-in circuits regardless of their antigen specificity. If it is the case, the model predicts the natural dominance of suppression over the induction under the physiologic condition. When an antetopic Th is activated by any antigen on APC, it can help only B cell clones displaying the relevant antigen in context of prototope (MHC-restricted cognate help). However, the antetopes on Th is recognized by Ts, and the signal is transmitted to the metatopic Ts. The Ts thus induced can suppress the antetopic cells in general, not only Th stimulated by the antigen. This leads to the suppression of the B cell response to this antigen but also to the general suppression of T cell dependent B cell responses. I believe that this actually occurs. It can explain the apparent unresponsiveness of autoreactive T cell clones that do exist in the presence of autoantigens. Antigenic competition and long term maintenance of tolerance may also be explained by the dominant and continuous nonspecific suppression. The antibody response to given antigen is only a temporal escape from this dominant suppression (derepression) caused by the selective stimulation of an antetopic T cell clone by the antigenic sign on APC. Other clones are not signified by the antigen, and are kept suppressed. The counter signal for augmentation of antibody response especially in the absence of Ts activation can reverse the situation. Possible infidelity of interpretation of antetopes into metatopes during the generation of circuit in the thymus may participate in the expansion of T cell repertoire. These should be other subject matters of semiotic considerations.

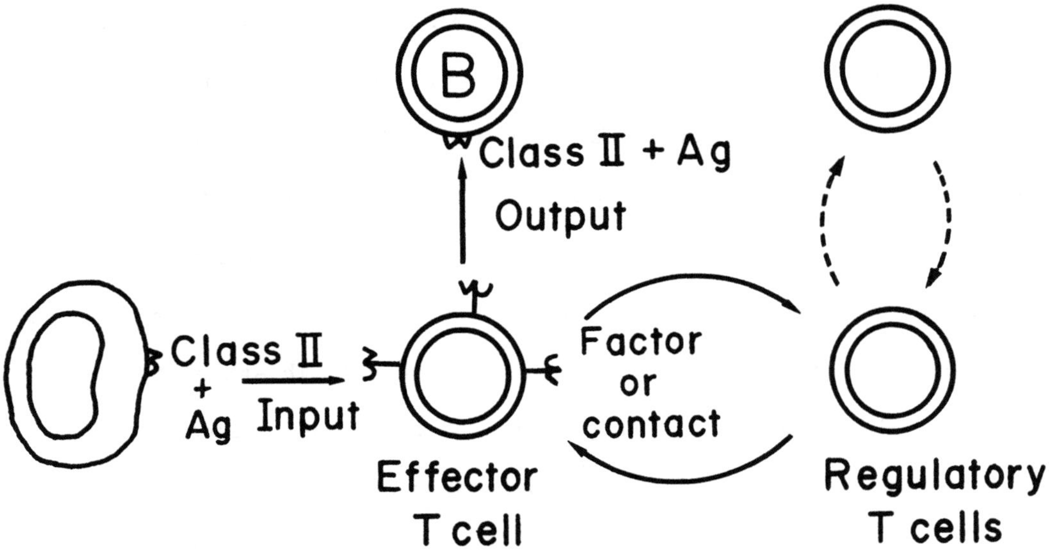

Figure 1. Minimal regulatory circuit activated by an antigenic input. Antigen is signified by the antigen-presenting cell to T cell which produces effector functions after perceiving the sign. An output is the helper effect for B cells having the same sign. This T cell interprets the initial sign and activites the second regulatory T cells that talk back to the effector T cell. Further interpretation and production of new signs are expected (dotted lines).

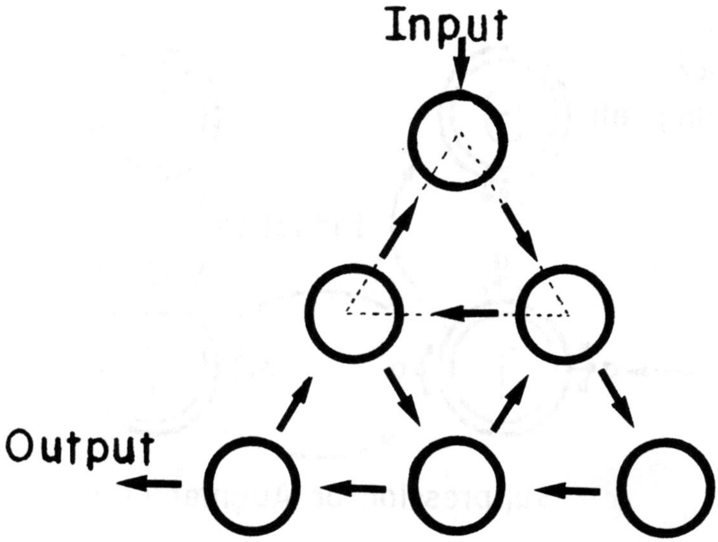

Figure 2. Triadic interactions between components of the immune system. The concept is based on premises stated in the text. There is an unidirectional flow of message which either produces an output or neutralizes the initial input within the system. The system is composed of interconnected triadic domains. The concept differs from the network which is based on the bidirectional flow of messages with two open ends (see text).

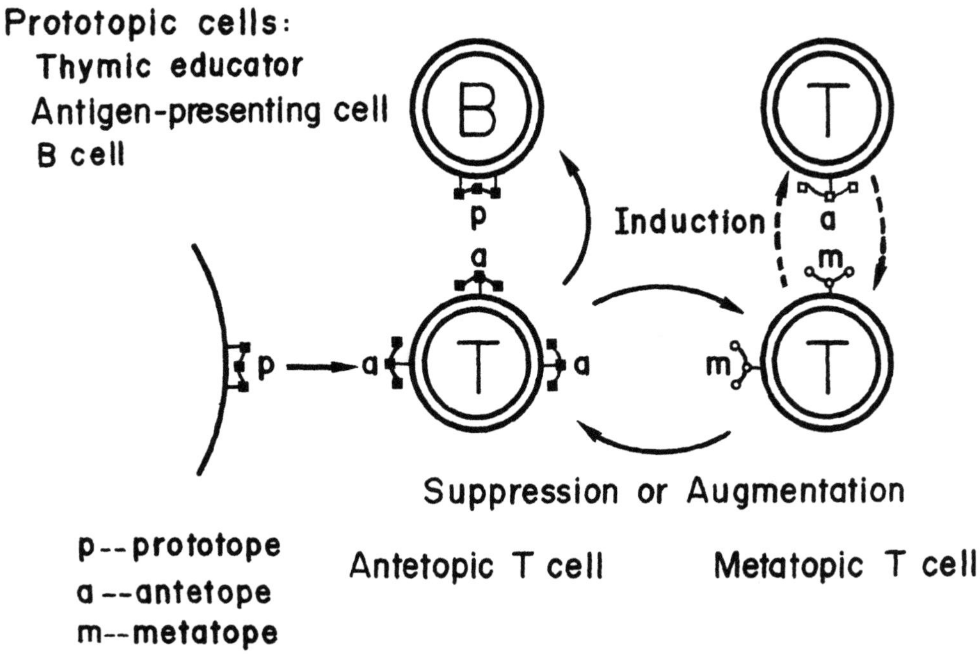

Figure 3. Construction of MHC-restricted circuits by different signs sequentially produced by the MHC isomorphism (see text).

Reference

1. Gershon, R.K.: Immunoregulation circa 1980: Some comments on the state of the art. J. Allergy clin. Immunol. 66:18, 1980.
2. Herzenberg, L.A., Black, S.J. and Herzenberg, L.A.: Regulatory circuit and antibody responses. Eur. J. Immunol. 10:1, 1980.
3. Tada, T., Abe, R., Uracz, W., Yagi, J., Miyatani, S., Kumagai, Y., Ochi, A. and Fujisawa, I.: Synaptic structures in the immunological circuit. In: Regulation of the Immune System. Cantor, H., Chess, L. and Sercarz, E. (eds.). Alan R. Liss, Inc., New York. p. 663, 1984.
4. Green, D.R., Flood, P.M. and Gershon, R.K.: Immunoregulatory T-cell pathways. Ann. Rev. Immunol. 1:439, 1983.
5. Dorf, M.E., Minami, M., Usui, M. and Aoki, I.: The NP suppressor-cell cascade. Progr. Immunol. V: 779, 1984.
6. Benacerraf, B.: Genetic control of the specificity of T lymphocytes and their regulatory products. Progr. Immunol. IV:419, 1980.
7. Klein, J.: In: Natural History of the Major Histocompatibility Complex. John Wiley & Sons, New York. 1986.
8. Sano, K., Fujisawa, I., Abe, R., Asano, Y. and Tada, T.: Major histocompatibility complex (MHC)-restricted minimal regulatory circuit initiated by a class II-autoreactive T cell clone. J. Exp. Med. 1987 (in press).
9. Nakayama, T., Kubo, M., Gomi, H., Asano, Y. and Tada, T.: MHC-restricted augmenting T cells induced by in vitro cultivation of antigen-preimed T cells. — A new parallelism between suppressor and

augmenting circuits —. (submitted).

10. Uracz, W., Asano, Y., Abe, R. and Tada, T.: I-J epitopes are adaptively acquired by T cells differentiated in the chimaeric condition. Nature 316:741, 1985.

11. Abe, R., Yagi, J., Asano, Y. and Tada, T.: Epitopes associated with the MHC restriction site of T cells. I. Selective expression of Iat epitopes on H-2-restricted helper T cells. J. Immunol. 1987 (in press).

12. Asano, Y. and Tada, T.: Epitopes associated with the MHC restriction site of T cells. II. Somatic generation of Iat epitopes on T cells in radiation bone marrow chimeras. J. Exp. Med. 165:87, 1987.

STRUCTURE AND DIRECTION OF MESSAGE IN THE IMMUNE NETWORK AND CIRCUITS

Constantin A. Bona
Department of Microbiology
Mount Sinai School of Medicine
New York, NY 10029/USA

> Certains déclarent qu'ils ont trouvé la vérité, d'autres proclament que son apprehension est impossible et d'autres enfin poursuivent leur recherche.
> Sextus Empiricus Hypotyposes I I-4

The major function of the immune system is the maintenance of macromolecular and cellular homeostasis in the body. Altered self and foreign macromolecules represent an aggressive challenge for the host because they have a tendency to alter the structure of messages which ensure the maintenance of homeostasis. The results of these alterations lead to the response against self-antigens (autoimmunity) or to foreign antigens (immune defense reactions).

The maintenance of macromolecular homeostasis representing the "steady state" of the system as well as an immune response elicited by self or foreign antigens is based on the ability of the immune system to perceive signs and to integrate them in a structural message. Since the immune system deals with signs having a precise meaning for the elements which compose it ("signifiers" and "signified"), it is worthwhile establishing whether or not the rules of semiotics can help the immunologists in understanding how the immune system works.

There are two different fields that we should distinguish:

a) Semiotics of the immunological dictionary, the glossary that each student finds at the beginning of an Immunology text book and which is continuously enriched by new terms invented to label new findings. These words are created by the human mind to have a precise meaning and consequently have an important anthropomorphic input.

b) Semiotics of the immune system, or immunosemiotics which could be designated as a new science of "immunosemiotics" if it would help us to understand the categories of "signotopes" (immunological signs perceived within the immune system) and their traffic, i.e. their integration into a message. This paper deals solely with the second aspect.

NATO ASI Series, Vol. H23
The Semiotics of Cellular Communication in the Immune System
Edited by E. E. Sercarz et al.

I believe that an immunosemiotic approach could be fruitful if we keep a few trivial facts in mind. First, there is a striking distinction between the signotopes of immunosemiotics and the "signs" used by semiotics of immunology. While the signs are created by the human mind and have a meaning, the signotopes do not have meaning in anthropomorphic terms because they are generated, translated and signified by objective rules. The signotopes belong to the domain of eternal truths. Secondly, the immune system has no consciousness. Immunosemiotics should not blindly borrow the rules and terminology of linguistics; while immunosemiotics deals with "infrastructural" phenomena governed by objective rules, semiotics deals with "suprastructural" phenomena. The rules and the terminology of linguistics are valuable models in our endeavor to interpret our observation; they should be shaped, examined and reexamined during our observations. Third, the structure of the immunosemiotic message and most importantly, the direction of the traffic of signs, is very different in the steady state (i.e. idiotype network), whereas the immune response is composed of immune circuits characteristic for each signotope.

The aim of this paper is to analyze the categories of signotopes and the unique features of the structure of the messages in these two different situations. I took as a model of a structural message, one commonly used in the semiotic science (1).

source -transmitter -signal -channel -signal -receiver -message -destinee
(interpreter) (interpretant)

Pierce's triad is composed of three types of signs - symbols, indices and icons. Obviously, immunosemiotics uses only two: indices physically connected with the object (antigen, lymphokines, etc.), and icons - similar to the objects (molecular mimicry such as antibodies carrying internal images).

In the communication process within the immune system an interpreter can translate one term into another language. An example of this are the antigen, presenting cells which are able to take up and process an antigen, and then present it in association with MHC self antigens to T lymphocytes.

The message in immunosemiotics is rather like a text since it represents a set of signals which are connected. As in language, one sign can be replaced by another and therefore can induce a set of different responses. The major characteristic of the interpretant (i.e. the lymphocyte) is that it is able to specifically select a definitive information input because of the specificity of its receptor which is encoded by a genetic program resulting from precise genetic events. The same interpretant can develop a response to a determined sign but also to many other signs because there are signs which contain a common text. Indeed, there are antigens of various origins which share cross-reactive epitopes.

In contrast to other somatic cells, B lymphocytes have a very specialized function in the immune system, namely, the ability to synthesize antibodies. The triggering of molecular events leading to the activation of resting B cells can be achieved in various ways since they bear on their surface a multitude of receptors able to interact with a variety of legends (Fig. 1). Some of these signals (effectotopes) are identical for all the clones in the same species or various species, induce direct response and represent type of stimulis-response interaction. The stimuli which are variable like the antigenic determinants (epitopes), or idiotopes of immunoglobulins induce a specific response (signotopes) by stimulating a single or a discrete number of clones.

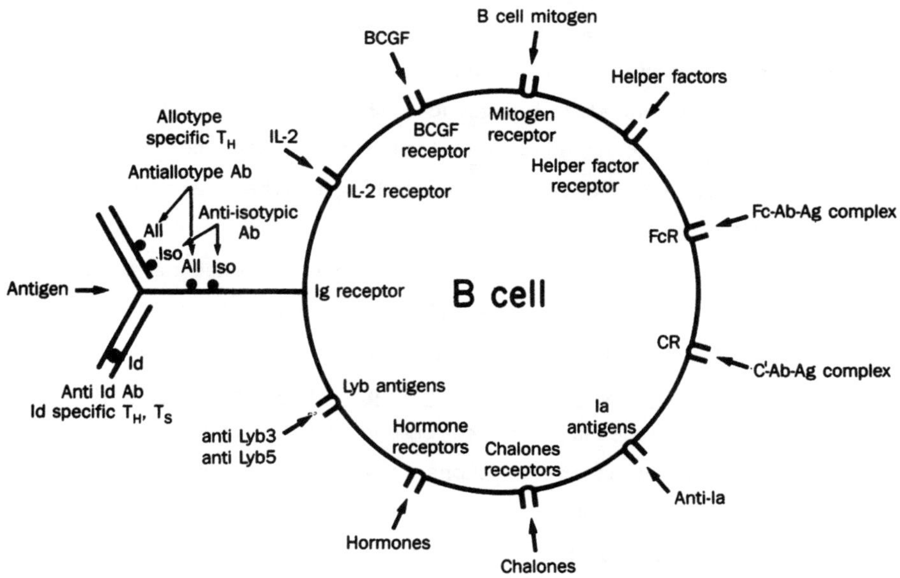

LEGEND TO FIGURE 1. Multitude of receptors able to recognize various signotopes from B cell environment.

Thus, a B cell specific for a given antigen such as β2-6 fructosan by virtue of its displaying a membrane immunoglobulin receptor which derives from the preferential rearrangement during the cell's development of only two germline genes (V_H441-4 and V_K10), can be activated by various signals interacting with many kinds of receptors associated with its membrane:

- signotopes recognized by the membrane Ig receptor, which can include, e.g., the β2-6 fructosan antigen ("index") or anti-A48Id antibody carrying the internal image of the antigen ("icon").
- effectotopes interacting with the membrane with Ig receptor (anti-allotype and anti-isotype antibodies), Fc receptor (Igs which bound levan or another antigen), complement (C) receptor (which bound C_3 resulting from the activation of C cascade subsequent to the binding of levan or another antigen), mitogenic receptor which bound levan or a completely different polyclonal activators, etc.

In spite of a broad ability of a single B cell to respond to a variety of signotopes, the B cell <u>is unable</u> to choose among various signals. When a B cell is simultaneously exposed to several stimuli, probably that with higher affinity for a given receptor induces activation. The flexibility of the immune system lies in its potential to recognize and respond to various signotopes without actively choosing so.

If a choice is apparent at first glance, then it is certainly governed by very specific rules based on physiochemical interactions between the shapes and chemical groups involved in the binding of the signotope (ligand) to the receptor of B cells (interpretant).

The flexibility of lymphocytes which could be envisioned as an active choice is actually limited by the number of receptors displayed on a single lymphocyte and the presence of a signotope in a given time.

1. <u>Direction of the communication of the message in the steady state</u>

In 1974, Jerne (2) proposed that the immune system might be viewed as a web of V domains based on interactions between idiotypes expressed on the receptor of B and T lymphocytes.

Since its first formulation, the network theory of regulation of the humoral response has been extensively remodeled through experiment (data reviewed in 4). However, its most fundamental concept, the recognition and regulation of self antibody idiotopes (antigenic determinants of antibody variable (V) regions) by lymphocytes, remains firm. An abundance of data, from a wide variety of experimental systems, have argued for its inclusion in any paradigm of immunology.

The idiotypic system actually represents an internal dictionary used by lymphocytic clones to communicate, to speak one to the other. The idiotype dictionary which requires the contribution of different genes (V,D,J, in the case of heavy chain; V,J in the case of the light and, in certain cases, even constant region genes (3)) might be very large and probably exceeds the library of the receptor of lymphocytes (paratope).

Jerne proposed that in the steady state every member of the system interacts with every other member and mutually regulates their levels of expression. According to the network theory for each member of the system able to interact with a foreign or self signotope operationally called Ab_1, which bears a particular idiotype (Id_1), there is a complementary clone called an anti-idiotype or Ab_2 which expresses its own idiotope (Id_2). This Ab_2 can induce the synthesis of Ab_3, etc (Figure 2).

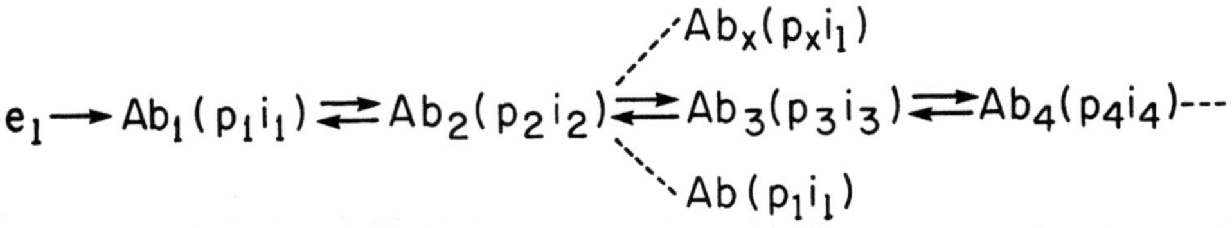

LEGEND TO FIGURE 2. Idiotype pathway (p=paratope, combining site; e=epitope, antigenic determinant, i-idiotope).

With this scheme in mind, it seems that suppression should be the basic regulatory pattern of the network since it is reasonable to assert that the suppression is the only mechanism which can guarantee the maintenance of a steady state. This steady state is assumed to maintain idiotype concentration at a low level in the absence of an antigenic challenge.

There is a perfect symmetry in the B and T cell compartments with respect to cells which can be classified as Ab_2, (Figure 3).

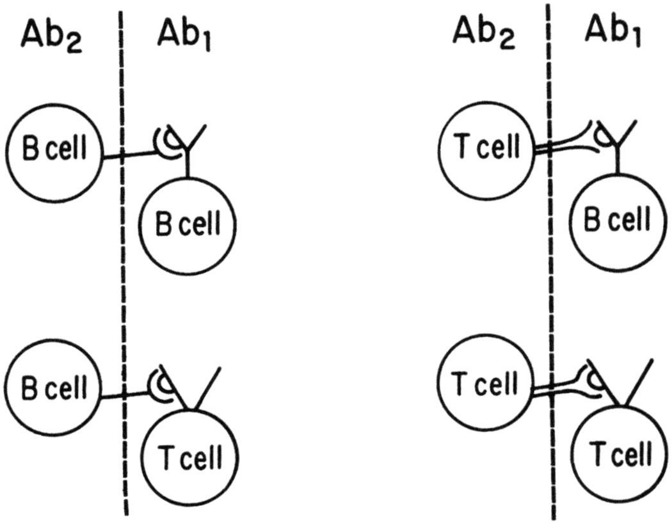

LEGEND TO FIGURE 3. Symmetry of cells bearing receptors recognizing idiotopes in T and B cell compartments (Ab1=cell recognizing foreign or autoantigens, Ab_2=cell recognizing idiotopes).

There are B cells which can produce antibodies recognizing the idiotypes of other B cells or T cells which recognize the idiotopes of the receptor of B cells or other T cells (data reviewed by Bona (4)).

The cells operationally defined as Ab_2 (anti-idiotypic) can be classified into three groups.

a) $Ab_{2\alpha}$ which are antibodies recognizing idiotypes of an immunoreceptor associated which the paratope or the framework of the receptor of other lymphocyte.

$Ab_{2\alpha}$ can induce either suppression or enhancement of Ab_1 clones depending on the concentration of Ab_2 in the system.
b) $Ab_{2\beta}$ are antibodies which carry the internal image of the antigen. The idea that each paratope recognizes an idiotope obviously requires that foreign signotopes (antigens) must cross-react with self signotopes (idiotopes). The idiotopes of an $Ab_{2\beta}$ are a typical iconic type of signotope since they mimic a foreign or self signotope which is similar but not identical with the idiotype of $Ab_{2\beta}$.

There are numerous examples demonstrating that anti-Id antibodies carrying an internal image mimic viral, bacterial and parasitic antigens, tumor antigens, self antigens or biologically active substances such as hormones, vitamins, etc. These types of antibodies actually connect the immune system with the outside world (foreign antigenic dictionary), as well as with the inside world (non-lymphoid somatic cells).
c) $Ab_{2\varepsilon}$-epibodies, a category of Ab_2 cells which recognize the idiotopes of an Ab_1 cell as well as the antigen interacting with the initial Ab_1 (data reviewed by Bona (3)) (Figure 4).

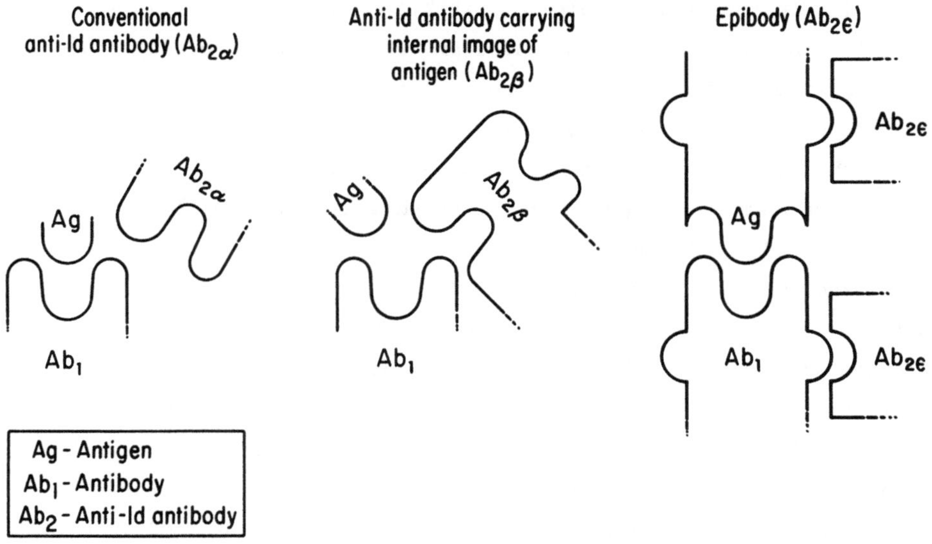

LEGEND FOR FIGURE 4. Various types of anti-idiotypic clones.

This category of Ab_1 appears as superantibodies (5) with a high connectivity since they can interact with the idiotopes of an Ab_1 as well as with the antigenic determinants of self or foreign antigens.

It appears that the flow of the information within the immune network in the steady state is multidirectional ("toutes azimutes").

This idea is strongly supported by two important findings:
a) the same idiotope can be shared by antibodies with various specificities (6). Indeed, in our laboratory we have shown that antibodies with various specificities for immunodominant sugars of LPS of gram negative bacteria (7) hemagglutinin of various influenza type A virus (H_1 and H_3) (8), various polysaccharides (β2-6 fructosan β1-6D galactan) (9) can share idiotopes.

Furthermore, Victor-Kobrin et al. (9) showed extensive idiotypic cross-reactivity among antibodies with various specificities encoded by the V_H J558 gene family and Monestier et al. (10) demonstrated that autoantibodies with various specificities (thyroglobulin, DNA, myelin basic protein, Sm, RFs) encoded by V_H J558 gene family share cross-reactive idiotopes of autoantibodies specific for DNA, Sm or RFs and even with antibodies specific for foreign antigens (influenza virus hemagglutinin, arsonate, GAT, etc).
b) antibodies in the steady state can exhibit multispecific binding properties. This contradicts the major paradigm that one paratope recognizes (can be signified) by a single epitope. A high frequency of multispecific antibodies were found in the neonatal state, as well as among autoantibodies which are encoded by genes which are similar (unmutated) or very closely related to the germline genes. This represents an important finding in the understanding of the activation of autoreactive clones because we do not know whether or not the forces (signotopes) which drive their expansion are the autoantigens or xenogenous antigens.

In the idiotype network, not all idiotopes are equal. We defined as regulatory idiotopes (11) those which play an important role in the regulation of the expansion of clones (privileged category of internal signotopes). They are characterized by several features, i.e. they are: a) markers of germline genes; b) autoimmunogenic; c) recognized by regulatory T cells and have the potential to become dominant; d) shared by antibodies with various specificities.

The analysis of antibodies produced by hybridomas obtained from animals injected with anti-A48Id antibodies and then immunized with bacterial levan showed that the majority of them used genes derived from V_H441-4 and V_K10 germline genes just as the A48Id-bearing ABPC48 myeloma protein which binds to the levan antigen.

However, an important fraction of these antibodies exhibit various specificities including the binding to other antigens such as inulin or galactan. This observation points out the role of the connectivity between individual clones signified by different antigens and a fundamental aspect of network regulation which we have designated as cross-regulation (9). We define as cross-regulation processes, including cellular and humoral mechanisms through which the self-recognition of regulatory idiotopes controls discrete populations of clones that display a variety of diverse antigenic binding specificities.

In summary, in the steady state, the idiotype dictionary ensures a high degree of connectivity between the individual clones composing the immune system, as well as a connectivity with the foreign and self-antigen dictionary.

A given clone can participate in multiple interactions since the idiotopes can be shared by clones with various antigen specificities or can mimic the foreign or self antigens.

2. Direction of communication during the immune response and within immunological circuits

The immune network theory views the immune system as a dynamic system. In the transition state which represents the immune response, the equilibrium between clones (Ab_1 and Ab_2) is upset subsequent to antigen stimulation. The transition state is determined by the clonal elements of the steady state. A new stable steady state is established after the extinction of the immune response. This new postimmune stable state is qualitatively different from the preimmune steady state due to the presence of memory cells (12) (Figure 5).

One of the major characteristics of the signotopes which drive the clones into the transitional state is that they actually represent to opposing messages: one immunostimulatory eliciting a humoral or cellular immune response and the other immunosuppressive which restricts the proliferation of clones. Depending on the dose of the antigen and other factors (age of the animal, route of the administration of antigen, irradiation, cyclophosphamide, etc.), the stimulus imparted by the signotope can be manifested either as a positive response or unresponsiveness. Thus, the signotope (antigen) has a dual function since it delivers both immunostimulatory or immunosuppressive signals. This function is related to an intrinsic property of the immune system which is the maintenance of homeostasis. An unlimited expansion of a normal clone could lead to an overwhelming of the system by a single clone, a situation that happens only in the case of aberrant differentiation of lymphocytic clones subsequent to an oncogenic event (myeloma, lymphoma, leukemia).

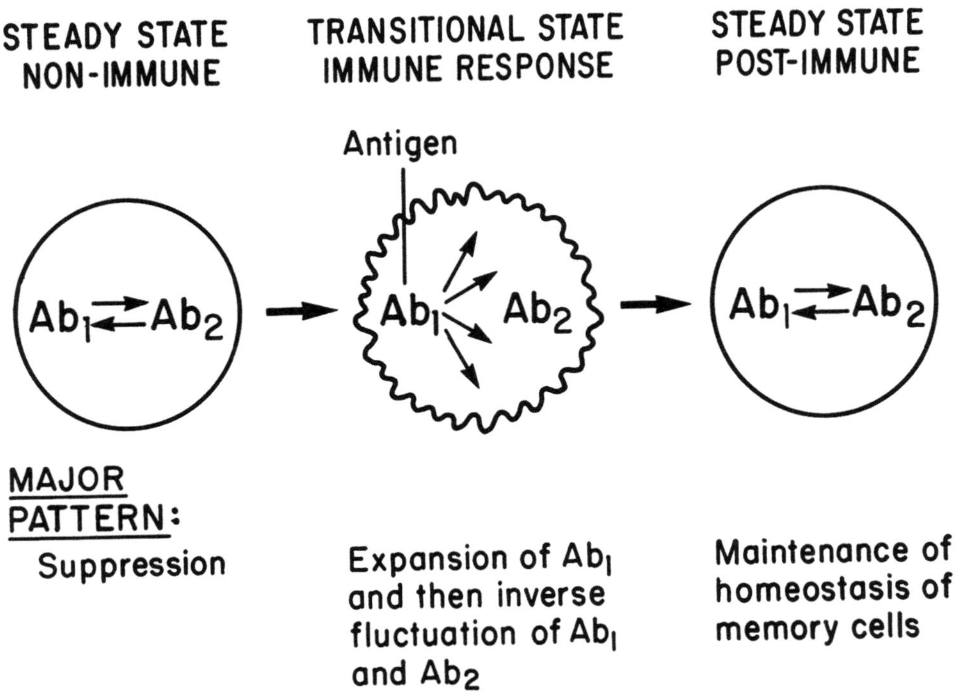

LEGEND FOR FIGURE 5. A dynamic view of the immune system based on idiotype regulation.

The concept of the dualism of signal delivered by a signotope clearly shows that the message utilized within the immune system does, in fact, comprise a text.

The syntactic rules of the text differ widely in function according to the nature of antigen as well as the cells involved in a given response which constitutes the immunocircuit. The components of a given immunocircuit depends on the nature of the antigen and on the imagination of the investigator used to rigorously define various elements which composed it.

The earliest model immunocircuits originating from studies devoted to understanding the cellular mechanisms of allotype suppression (13) or the extinction of the immune response (14) were beautifully simple. A B cell specific for a given antigen is expanded either by antigen specific or allotype specific helper T cells.

This expansion is inhibited by antigen specific or allotype specific suppressor cells exerting their effect on helper T cells. Tada et al. (15) introduced other interpreter-interpretant elements into immunological circuits when he defined new cells such as inducer T cells, transducer T cells, etc. The research groups headed by Sercarz and Benacerraf (16,17) showed the complexity of the text of the message within immunocircuits.

The immunologic circuits (each probably characteristic for a given antigen) become activated subsequent to the recognition of the antigen by a lymphocyte genetically preprogrammed to recognize it, and they are connected with other elements which can recognize different signotopes.

These circuits have two major characteristics. a) the traffic of the signotope is <u>unidirectional</u> and circular. b) the text <u>containing</u> the signotope is <u>constantly altered</u>. Independent of its positive or negative feedback regulatory pattern of an immune circuit, the traffic is unidirectional because the antigenic stimulus triggers a series of molecular mechanisms sequentially activated in various cells.

The structure of the message containing a given signotope is permanently altered because of the individual contribution of a single element of a circuit which functions as interpreter when it reads the text and as interpretant when it delivers the text.

<u>Example</u>

<u>Macrophage</u>	<u>T inducer</u>	<u>T suppressor</u>
reads antigen	reads	reads
and	Antigen + MHC	Ts_1
translates into	and translates into	and translates into
Antigen + MHC	a lymphokine (Ts_1)	Ts factor, etc.

In conclusion, it appears that there are important differences between the idiotype network which characterizes the interactions within a steady state and immune circuits which characterize interactions during immune responses with respect to the traffic of signotopes, and the complexity of the message used in lymphocytic communications.

Acknowledgements

This work was supported by National Science Foundation grant. I am grateful to Drs. Th. von Uexkull, J.L. Schulman, E. Sercarz and F. Celada for reading this manuscript and for their suggestions.

Literature references

1. Eco, U. A theory of semiotics. Indiana Univ. Press, Bloomington, London, 1979.
2. Jerne, N.K. (1974). Towards a new theory of the immune response, Ann. Immunol. (Paris) 125 C 373-389.
3. Cronkhite, R.I. and Cerny, J. (1986). A novel idiotypic determinant on phosphorylcholine-binding immunoglobulins restricted isotype and allotype. J. Immunology 136, 3729-3733.
4. Bona, C. Regulatory Idiotopes. J. Wiley & Sons, New York, NY (1987).
5. Dweyer, S.D. Idiotype connectivity of antibody responses specific for self and nonself antigens in Biological Applications of Anti-idiotypes (C. Bona, ed). CRC Press, Boca Raton, FL. (1987).
6. Oudin, J. and Cazenave, P.A. (1971). Similar idiotypic specificities in immunoglobulin fractions with different antibody functions or even without detectable antibody function. Proc.Natl.Acad.Sci.USA, 68, 2616-2620.
7. Hiernaux, J., and Bona, C.A. (1982). Shared idiotopes among monoclonal antibodies specific for different immunodominant sugars of lipopoly-saccharide of different gram negative bacteria. Proc.Natl.Acad.Sci.-USA.,79, 1616-1620.
8. Moran, T., Liu, Y.N.C., Schulman, J., and Bona, C. (1984). Shared idiotopes among monoclonal antibodies specific for A/PR/8/34 (H1N1) and X31 (H3N2) influenza viruses. Proc.Natl.Acad.Sci.USA 81, 1809-1812.
9. Victor-Kobrin, C., Bonilla, F.A., Bellon, B., and Bona, C.A. (1985). Immunochemical and molecular characterization of regulatory idiotopes expressed by monoclonal antibodies exhibiting or lacking β2-6 fructosan binding activity. J.Exp.Med. 162, 647-66.
10. Monestier, M., Bonin, B., Talal, N., Dang, H., Maurer, P., Rose, N., Datta, S., and Bona, C. Binding to foreign antigens and sharing of idiotopes among autoantibodies with various specificities encoded by V_H J558 gene family. (in preparation).
11. Bona, C.A., Heber-Katz, E., and Paul, W.E. (1981). Idiotype-antiidiotype regulation I. Immunization with a levan-binding myeloma protein leads to the appearance of auto-anti (anti-idiotype) antibodies and to the activation of silent clones. J.Exp.Med. 153, 951.
12. Bonilla, F.A., and Bona, C.A. Idiotypic immunoregulation: interclonal connections based on a special category of idiotopes, in Idiotypes. (M. Reichlin and D.J. Capra, eds). Academic Press, New York 1987.
13. Gershon, R.K. T cell regulation: The second law of thymodynamics, in The Immune System: genes, receptors, signals (E.E. Sercarz, A.R. Williamson, C.F. Fox, eds) Academic Press, New York 1974, p471.
14. Herzenberg, L.A., Black, S.J., and Herzenberg, L.A. (1980). Regulatory circuits and antibody response. Eur. J. Immunol. 10, 1-7.
15. Tada, T., Okumura, K., Miyatani, S., Ochi, A., Nanaguchi, H., and Karasuyama, H. (1983). Restricting elements in the immunological circuitry: the role of I region-controlled determinants in immune networks (C. Bona and H. Kohler, eds). Ann. New York, Acad. Sci. 418, 189-197.
16. Sercarz, E.E., Araneo, B., Benjamin, C.D., Harvey, M., Metzger, D., Miller, A., Wicker, L., and Yowell, R. (1983). The design of regulatory circuitry: predominant idiotype and the idea of regulatory parsimony in Immune network (C. Bona and H. Kohler, eds). Ann. New York Acad. Sci. 418, 198-205.
17. Germain, R.N., Sy, M.S., Rock, K., Dietz, M.H., Green, M.I., Nisonoff, A., Weinberger, J.L., Ju, S-T., Dorf, M.E., and Benacerraf, B. (1981). The role of idiotype and the MHC in suppressor T cell pathways in Immunoglobulin idiotypes (C. Janeway, E.E. Sercarz, and H. Wigzell, eds). Academic Press, p 709.

Modes of Communication Within the Immune System: Action or Reaction?

Patrick M. Flood, Ph.D.
Laboratory of Cellular Immunology
Howard Hughes Medical Institute
Department of Pathology
Yale University School of Medicine
New Haven, CT 06510

The immune system is a complex intertwining of individual components who collectively have one major purpose in mind: to recognize and eliminate antigens which are perceived by the immune system to be "different," i.e. nonself. To do this, and to prevent the immune system from launching an attack on itself which may prove fatal, the immune system has developed a complex mechanism of regulation which functions remarkably well (most of the time). This system of regulation is based on a system of cellular co-recognition and communication that theories of immunosemiotics are based.

Three of the main aspects of semiotics which need be addressed in the construction of immunosemiotic theory are signs, recognition, and response. Given the enormous diversity of cellular components within the immune system it is imperative that a system is created in which an unambiguous sign is delivered in quick and efficient fashion to the appropriate target, resulting in the elicitation of the desired response. To do this, the immune system has devised specific targetting mechanisms, utilizing both antigen-specific and antigen-nonspecific activities, each of which represents its own set of problems that must be dealt with in a well functioning communication system, and each of which has a unique action on its target population depending on the receptivity of that cell to that particular signal.

NATO ASI Series, Vol. H23
The Semiotics of Cellular Communication in the Immune System
Edited by E. E. Sercarz et al.

The importance of information trafficking is the cornerstone of all of immunology, since few if any immune responses proceed without the interaction of functionally distinct cell types. These interactions often involve very complex circuits in which a number of cell types must interact in a well-defined pathway of paired sets which ultimately leads to an immune activity. Information transfer among regulatory T cell subsets is mediated by biologically active factors consisting of two molecules: one that binds antigen, and another that expresses MHC-linked determinants such as I-J or $I\text{-}A_t$ (1). It appears that these products exhibit four major properties which are of interest from an immunosemiotic perspective: 1) these factors are antigen-specific, in that they only regulate the immune responses of cells responding to the appropriate, but not the inappropriate, antigen; 2) these factors are genetically restricted, in that they only regulate the immune responses of cells which exhibit the same genetic polymorphisms at MHC or Igh-V that are exhibited by the cells which secrete the factors; 3) these factors deliver a reproducible biological message to their target cell, in that a suppressor-effector factor always suppresses its target cell, while a suppressor-inducer always induces the regulatory pathway; and 4) these factors always deliver their message to the appropriate target cell within the regulatory circuit, in that suppressor-inducer factors always deliver their message to suppressor-transducer cells, while suppressor-effector cells always deliver their message to T helper cells. Therefore, in harkening back to our three aspects of immunosemiotics that must be addressed by the immunologist, the "sign" used by the immune system, i.e. the regulatory factor, appears to be consistently recognized by the appropriate target cell and the response is appropriate.

While the problems posed to the immune system by antigen-specific signals are indeed immense, the problem of antigen-nonspecific information trafficking is even more complicated. In trying to unravel the complexities of imformation trafficking within the immune system, the immunologist is faced with the problem that while most if not all immune responses are antigen specific, the activation and growth of lymphocytes and monocytes is accomplished primarily by secreted antigen-nonspecific lymphokines. Therefore, one must recognize that in order for a cell to interact with another via nonspecific mediators, a very important part of that interaction must deal with the delivery of that "sign" only to the desired target. In dealing with these problems of targetting, the immune system must first find a way to deliver an antigen-nonspecific biological message to the desired antigen-reactive recipient. Relevant to this problem is the discovery that individual lymphokines such as IL-1, IL-2, IL-3, IL-4, IFN γ and GM-CSF have multiple activities depending on the target of the activity and their relative concentration of interaction with that target at any one particular time. For example, it has now been well established that IL-1 has effects on B cell function, fibroblast proliferation, muscle proteolysis, prostaglandin E_2 and collagenase production by rheumatoid synovial cells, hepatic acute phase protein synthesis, fever and slow wave sleep induction, ACTH regulation, and most importantly for us, T cell chemotaxis and activation (2). Many of these activities, either directly or indirectly, affect antigen specific immune responses. In addition to multiple target cell capabilities, two other problems compound the issue: multiple sources and multiple forms of IL-1. Sources of IL-1 include monocytes, keratinocytes, fibroblasts, B cells, T cells, and isolated calvarial cells among others (2-4); forms include a 17kD and a 33kD form which can exist as both IL-1α and IL-1β (and perhaps IL-1γ ?), all of which appear to work through the same cell surface receptor (4,5). Compounding this problem even further is the fact that many of the activities of IL-1 (e.g. stimulation of fibroblast production and release of

PGE_2 and collagenase, stimulation of bone resorption in vitro, activation of T cells, and activation of B cells), can be accomplished by other lymphokines such as tumor necrosis factor (TNF) α and β, platelet-derived growth factor (PDGF), fibroblast growth factor (FGF), epidermal growth factor (EGF), IL-2, IL-4, and GM-CSF (6-8). It is also quite clear that IL-1 can synergize with IL-2 or IL-4 to promote growth and differentiation of T cells depending on the time of administration of these lymphokines and their relative concentrations (9). Yet with all these variables, it appears as though immune responses initiated by lymphokine-producing cells and accomplished by antigen-reactive cells are: 1) antigen specific; 2) genetically restricted; 3) directional; and 4) consistent. Therefore, one can begin to appreciate the complexities facing the immunoregulatory semioticist in trying to decipher the code used by the immune system (and other systems as well) during the activation of an antigen-specific response in vivo.

How, then, does the immune system deal with the regulation of immune responses by secreted mediators when it appears as though any number of undefined cells or factors may influence the exact variable you are trying to quantify? To solve these questions we must turn outside ourselves to the semioticists who may provide us with two important lessons. The first lesson deals with context, the second with recognition. The semioticists tell us that first, the sign to be delivered must be as unambiguous as possible. That is, in order to consistently deliver the message that one desires, the message must be made simple and must not contain multiple options for the receiver. Second, delivery must be specific and the receiver must be able to recognize and decipher the message easily, in order not to allow the message to "fall into the wrong hands."

We have extensively studied the nature of information trafficking by antigen-specific regulatory cells, and a picture is emerging which may allow us to answer some of these questions. Our results suggest that the semiotic activity of regulatory factors can be broken down into its component parts, with each component being assigned to a particular functional domain of these factors (10,11). The first result suggests that the antigen-binding component of the regulatory factor, which we have termed the ABM, is the molecule which conveys antigen specificity to the factor upon its interaction with its target cell. It is likely that antigen serves primarily to increase the likelihood of encounter between antigen-specific factors and antigen-specific cells by binding to antigen and being sequestered into areas of high antigen concentration, which are also places where antigen-reactive cells tend to accumulate. In this way, specificity is achieved by a mutual attraction toward the common ligand which drives both the inducer and the regulatory cell.

The second result suggests that the MHC-bearing component of the regulatory factor, which we have termed I-J, is the molecule which conveys genetic restriction to the factor upon its interaction with its target cell. This genetic restriction is linked primarily to the Igh-V gene polymorphism, and is a learned restriction between the factor and its target cell (9). We have found that the target cell of factor activity has receptors for this molecule of the factor, but not for the antigen-binding chain of the factor. Although the exact details of how Igh-V linked determinants play a major role in recognition of regulatory factors is not known, the binding of this molecule is required for the biologic activity of this factor to be exhibited, and this binding shows the same genetic restriction as does the factor's activity on its target cell population. It therefore appears that while the ABM molecule is totally responsible for antigen localization and antigen specificity of the factor, interaction of the factors with their target cells is accomplished primarily by the interaction of MHC-bearing molecules with "self" receptors on target cells which act as a sign that the cell is about to be regulated.

How do these factors know which cells are "self" so as to deliver their message to the appropriate one within the regulatory circuits? In order to address this question we have made a basic assumption: the difference in biological activity between different regulatory molecules must be reflected by differences in the molecular structure of the factor. These differences could be manifest as differences in primary amino acid sequence, glycosylation differences, differences in associated proteins, or differences in tertiary structure. If so, these differences may be detectable by serologic analysis of different factor-specific monoclonal antibodies, or by differences in binding to target cell receptors. The first relevant observation was that it appears that target cells for factor activity only have receptors for the MHC-bearing molecule. The second observation was that, by absorption studies, T helper cell clones only bind to and remove the MHC-bearing molecule from factors which regulate T helper cell activity (e.g. suppressor-effector factor), while the MHC-bearing molecule of suppressor-inducer factor is only bound by T suppressor-transducer cells, the physiologic target of this factor. The third observation was that serologic analysis of factors with different target cell specificity suggests that the MHC-linked determinant (I-J) on this MHC-bearing molecule exhibits heterogeneity which correlates with target cell specificity. Therefore, by binding studies and serologic analysis, it appears that MHC-bearing molecules of regulatory factors mediate the T cell subset restriction exhibited by regulatory circuits.

The final question remaining about regulatory factors is the source of the biologic signal delivered by these factors; i.e. what is the molecular source of the sign that must be interpreted by the recipient? Since we determined that the MHC-bearing molecules determine subset specificity, we constructed "hybrid" factors, taking ABM molecules of a given functional activity (i.e. induction of suppression) and mixing it with a second MHC-bearing molecule of a different target specificity and asked the biologic consequences on an immune response. The finding was that whenever an ABM from a suppressor-effector factor was used, the biologic result was suppression regardless of the target cell specificity of the factor. In addition, whenever an ABM from a suppressor-inducer factor was used, the biologic result was induction, again regardless of the target cell specificity of the factor. Serologic analysis of a large panel of regulatory factors, both monoclonal and polyclonal, suggests that monoclonal anti-C_t reagents of Owen (12) and anti-TsF reagents of Ferguson et al. (13,14) delineate this inducer from suppressor phenotype on ABM from these factors. Therefore, by functional and serologic analysis it appears as though the ABM is the message that is received by the target cell.

I have just described a rather sketchy but intriguing story about how information between antigen-specific regulatory cells may be trafficked using antigen-specific biologic messages. However, as stated before, in the activation, growth, and differentiation of lymphocytes in response to antigen, messages directing these cells appear to be antigen nonspecific and very profuse in both their target cell specificity and in their biologic activity. How then does the immune system deliver to the appropriate antigen-reactive cell the appropriate biologic signal? Again, turning to the semioticists, we learn that a very good way of insuring the delivery of the appropriate message to the appropriate cell type is based on a system of co-recognition, verification, and interaction. Spies working for governmental agencies intent upon passing information only to those co-operatives to whom it was intended have very effectively used this type of system throughout history - primarily by agreeing on certain ground rules of recognition and then insuring that recognition by the passage of a signal which verifies the encounter. This is also how the immune system works to insure the passage of potentially valuable information only to the recipient who is designed to use it effectively.

Let us therefore lay down some ground rules for lymphocyte:lymphocyte interactions: 1) lymphocytes release their messages only after the appropriate stimulatory signal; this means that small resting antigen-specific cells do not constitutively produce lymphokines. 2) the activation of cells to produce lymphokines must follow some recognition event; this means that cells will not produce lymphokines unless told to do so by some very specific activation signal; 3) target cells must be able to decipher the message; that is they must have some way of using this message once it has been given to them; and 4) lymphocytes must have some way of insuring that only the desired individual uses the message. It doesn't make sense to devise an elaborate scheme for insuring specific recognition if anyone in the vicinity can use the message once the cell has been activated to deliver it.

Luckily for us, it appears as though lymphocytes have excellent mechanisms for insuring that all four of these criteria are met. In using the interaction of T and B cells as a model, T cell produce certain lymphokines which appear to augment or suppress the activity of antigen-reactive B cells. These lymphokines include IL-1, IL-2, IL-4, GM-CSF, and IL-5, which appear to augment B cell activity, and IFN-γ , TNF-β , and perforin, which suppress the activity of B cells. However, it appears that these lymphokines are produced by T cells only after the appropriate engagement of the T cell receptor (Ti/T3)/L3T4 complex (15), or after the engagement of certain other T cell activation antigens such as Thy-1 or Ly-6 (16,17). This engagement can be accomplished by, of course, Ag/MHC complexes on APC, or by monoclonal antibodies specific for either the Ti/T3 complex on T cells or for these other activation molecules. Other cells which produce these lymphokines also appear to be under some kind of regulation of gene expression, although the exact nature of that regulation is still very much a mystery. While the exact nature of the signalling which activates T cells through these receptors has yet to be determined, the fact is that T cells can release these lymphokines only after this signal has been delivered, insuring the appropriate control over the mismanagement of very valuable biological messages.

Since T cells must be activated to release lymphokines, the activation event can serve as the restricting element to insure proper activation. Once again, the interaction of T and B lymphocytes has served us well as a model to illustrate this point. Two characteristics of this interaction suggest that it is highly specific. First, the T cell must recognize processed antigen on the surface of the B cell through its Ti/T3 receptor; this means that only T cells specific for the processed antigen can be stimulated. In addition, it appears that B cells which are specific for a particular antigen can take up and process antigen much more efficiently than nonspecific B cells. Second, the T cell must recognize major histocompatibility complex (MHC) genes on the surface of the B cell in addition to processed antigen. While this does not insure that only the specific B cell gets activated, levels of MHC molecules on the surface of B cells can serve as another level of regulation to influence the outcome of the T:B cell interaction.

In order to decipher the message, cellular targets must, of course, be able to receive it. Therefore, target cells must express lymphokine receptors. Unfortunately, not much is known about the expression and regulation of lymphokine receptors, mostly because the nature of these receptors is only now being elucidated. It is likely, however, that co-recognition and activation events increase the expression of lymphokine receptors, in order to increase the sensitivity of responding cells to these messages. Preliminary results with the IL-2 and IL-4 receptors, in fact, suggest just that.

However, while increases in the number of lymphokine receptors may help to explain specific triggering, it is obvious that another explanation is also required. The simplest explanation is that cells only deliver their message to the appropriate cells by insuring there is only one message delivered. This is, in fact, not the case, since a large number of different lymphokines are produced by T cells activated in the aforementioned ways. How, then, can it be done? Perhaps by insuring that while a large number of different messages are produced, they only reach the intended target. It has recently been shown that a cytoskeletal polarity of the Ti/T3 complex develops in T cells when they recognize foreign antigen/MHC on the B cell surface (18). In addition, it has been found that the secretion of B cell activating lymphokines parallels the polarity found with the receptor (19). This indicates that activated T cells appear to <u>direct</u> the secretion of their lymphokines only toward the target cell they interact with, insuring perhaps that the target cell is the <u>only</u> one that receives the message.

The immune system has evolved to insure that only the appropriate antigen specific cells respond under most conditions. To do this, it has developed a rather sophisticated mechanism of cellular co-recognition and pairing, and the passage of information between these cells. We have just scratched the surface of how this information is sent, delivered, and received. In trying to learn the rules, one is often best served by seeing how other systems have accomplished similar goals. Semiotics, the science of information transfer within an interactive system, may ultimately prove to be beneficial or deleterious in our quest to find the rules for information transfer within immunologic systems. However, at least for those of us who shared some time with the semioticists trying to find out, it was a great place to start.

REFERENCES

1) Green, D., Flood, P.M., and Gershon, R.K. 1983. Ann Rev. Immunol. 1:439.

2) Dinarello, C.A. 1984 Rev. Infect. Dis. 6:52.

3) Kupper, T., Ballard, D., Flood, P.M. et al. 1986 J. Exp. Med. 164:2059.

4) Lomedico, P.T., Kilian, P.L., Gubler, U. et al. 1987 Cold Spring Harbor Symp. of Quan. Biol. 51: in press.

5) Mizel, S.B., Kilian, P.L., Lewis, J.C. et al. 1986 J. Immunol, in press.

6) Dukovich, M., Severin, J.M., White, S.J., et al. 1986 Clin. Immunol. Immunopath. 38:381.

7) Horowitz, M., Baron, R. and Vignery, A. 1984 Proc. Natl. Acad. Sci. USA 81:2818.

8) Pike, B.L. and Nossal, G.J.V. 1985 Proc. Natl. Acad. Sci. USA 82:8153.

9) Horowitz, M., Kupper, T.A., and Flood, P.M. 1987 J. Immunol., in press.

10) Flood, P.M., Chue, B., and Whitaker, R.B. 1985 J. Immunol. 135:933.

11) Flood, P.M. 1986 Mediators of Immune Regulation and Immunotherapy, (Singhal and Delovitch, eds.) Elsevier Science Publishing, New York, p81.

12) Owen, F. 1983 Adv. Immunol. 34:1.

13) Ferguson, T., Beamon, K.D., and Iverson, G.M. 1985 J. Immunol. 134:3163.

14) Ferguson, T. and Iverson, G.M. 1986 J. Immunol. 136:2896.

15) Haskins, K., Kappler, J. and Marrack, P. 1984. Ann. Rev. Immunol. 2:23.

16) Gunter, K.C., Malek, T.R., and Shevach, E. 1984 J. Exp. Med. 159:716.

17) Rock, K.L., Yeh, E., Gramm, C., et al. 1986 J. Exp. Med. 163:315.

18) Kupfer, A., Swain, S.L., Janeway, C.J. et al. 1986 Proc. Natl. Acad. Sci. USA 83:6080.

19) Poo, W-J., Conrad, P., and Janeway, C.A. 1987 submitted.

OF WORDS, GENES AND MUSIC

Susumu Ohno
Beckman Research Institute of The City of Hope
1450 East Duarte Road
Duarte, CA 91010

I pay homage to Umberto Eco who attended this meeting by quoting the very first sentence of PROLOGUE to his widely acclaimed book "THE NAME OF THE ROSE" in English translation: "In the beginning was the Word and the Word was with God and the Word was God". It would be noted that in this 17-word-long sentence, the Word re-currs thrice and God twice; all together, recurring words occupying half of the sentence. Indeed, the essence of any good writing appears to depend upon the recurrence at more or less regular intervals of the same or similar sounding words which give a lyrical quality to it. Julius Caesar's announcement to the Roman senate of his victory at Zela (47 B.C.) survives to this day, only because all three word-combinations uttered in succession began with v and ended in i; "*ve'ni, vi'di, vi'ci*". This then is the extreme of recurrence with no interval at all. In immunology, the antigen-specific cooperation between *helper* T-cells and antibody-producing B cells appears again to depend upon recurrence of the same word (the same signal). As schematically illustrated in Figure 1, a macrophage phagocytizes antigen "A" and digests it to a number of small peptide fragments. Of those digested fragments, an amphipathic α-helical fragment is preferentially chosen and presented to the outside world by the antigen-presenting macrophage in conjunction with the self class II MHC antigen (extreme left of Figure 1). A clone of T-cells which happen to possess the membrane-bound receptor that fits this antigen "A" α-helical fragment + self class II MHC antigen complex, now becomes antigen-specific *helper* T-cells (middle of Figure 1). But how this clone of T-cells can selectively recognize and help not a single clone but a number of clones of B-cells equipped with membrane-bound anti-antigen "A" anti-bodies of IgM and IgD types? For it would be recalled that antibodies are constructed to recognize antigens *per se* not complexed with self class II MHC antigen. Further-

NATO ASI Series, Vol. H23
The Semiotics of Cellular Communication in the Immune System
Edited by E.E. Sercarz et al.

more, each polypeptide antigen usually present not one but a number of antigenic determinants. Accordingly, there ought to be and are several different clones of B-cells; their antibodies being directed against different antigenic determinants of antigen "A". Nevertheless, having been endowed with membrane-bound anti-antigen "A" antibodies, all these different clones of B-cells would manage to concentrate antigen "A" on their cell surface. These antigen-antibody complexes shall be lysostripped off the plasma membrane and shall be digested inside B-cells. As the digestive enzyme involved is of the same sort as that present in the macrophage, antigen "A" is digested to the same variety of peptide fragments and the same amphipathic α-helical fragment is chosen and presented by B-cells to the outside world in conjunction with self class II MHC antigen. It is this complex which the receptor of *helper* T-cells recognizes, thus, resulting in antigen-specific help to expand all clones of anti-antigen "A" B-cells (extreme right of Figure 1).

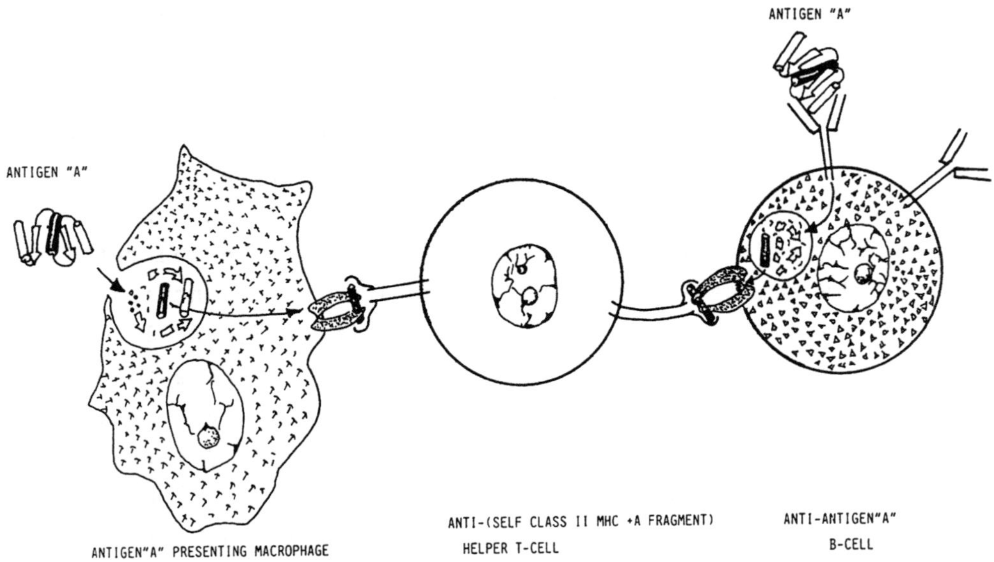

Figure 1

How *helper* T cells are able to provide the antigen specific help to multiple clones of B cells is schematically illustrated. This help is based upon antigen-presenting macrophages and antigen-specific clones of B cells uttering the same word which is perceived as such by membrane-bound receptor of *helper* T cells. At the extreme left, the specific antigen (antigen "A") is depicted as a polypeptide chain comprised of alternating α-helical and β-sheet forming segments; of those, one amphipathic α-helical segment is shown as a black barrel. A macrophage, at the left, phagocitizes antigen "A", and in a specific intracellular locality but not in a lysosome, antigen "A" is digested by a particular protease to several peptidic fragments. One amphipathic α-helical fragment then is preferentially chosen and presented to the outside world, complexed with self class II MHC antigen. A clone of T cells equipped with the receptor that fits this complex presented by the antigen-presenting macrophage now becomes anti-antigen "A" specific *helper* T cells as shown in the middle. On the other hand, still membrane-bound antibodies of dormant anti-antigen "A" B cells shown at the right recognize any of the antigenic determinants present on antigen "A" but never a complex formed between self class II MHC and antigen "A" aphipathic α-helical fragment. Yet they can be recipients of the help from anti-antigen "A" specific T cells, because a complex formed between antigen "A" and specific membrane-bound antibody is taken inside B cells by pynocytosis and subsequently antigen "A" is digested by the same protease as present in macrophages. Of digested fragments, the same amphipathic α-helical fragment is preferrentially chosen and presented by B cells to the outside world complexed with self class II MHC antigen. This enables anti-antigen "A" specific *helper* T cells to see the same word on the plasma membrane of anti-antigen "A" specific B cells as it had seen on the antigen-presenting macrophage plasma membrane; hence antigen-specific help to cause clonal expansion of the antibody secretion by anti-antigen "A" specific B cells.

All in all, it would thus appear that the antigen-specific cooperation between T-cells and B-cells is based upon one principle; that when confronted with the same sentence (antigen "A"), both antigen-presenting macrophage and anti-antigen "A" B-cells chose the same word (a particular amphipathic α-helical segment) out of that sentence and crowns it with the same adjective (self class II MHC antigen).

COMPLEMENTARY RECOGNITION IS BUT A FORM OF THE HOMOLOGOUS RECOGNITION

In immunology, one often speaks of specific antigen-antibody interactions as examples of recognition based upon the complementarity between two components. Nevertheless, the fact is that all components of the adaptive immune system are composed of strings of repeating units ultimately derived from the common ancestral unit. This unit commonly referred to as the β_2-microglobulin-like domain is made of 90 to 100 mostly hydrophobic amino acid residues, the relative abundance of hydrophilic Ser and/or Thr also being a conspicuous feature. These residues are folded into three to five loops of anti-parallel β-sheet forming segments. Contacts between neighboring β-sheet forming segments are maintained through hydrogen bonds mostly formed between Thr-Thr, Thr-Ser or Ser-Ser, and the whole structure is compacted by the presence of one intradomain disulfide bridge.

It now appears that the immediate ancestor of genes for the adaptive immune system was CAM (cell adhesive molecule) gene engaged in organogenesis of early embryos. In the extracellular portion of N-CAM specific for neuronal organization, four successive β_2-microglobulin-like domains were found (Hemperly et al.,1986). Through these domains, N-CAM engages in homologous recognition, thus, aggregating similar neuronal cells; the first step in neuronal organization. It is fitting that all components of the adaptive immune system evolved from CAM; the original mediator of cell-cell interaction. The point to be made here is that the β_2-microglobulin-like domain originally evolved to engage in homologous, not complementary, recognition. Accordingly, recognition of class I and Class II MHC antigens by T-cell receptors and B-cell antibodies, as well as that of idiotypes by another T-cell receptors and B-cell antibodies are homologous recognition *sensu stricto* ; the notion of complementary recognition being more of an illusion than reality

SEARCH FOR THE ULTIMATE ANCESTOR

Implicit in the above stated notion that the immediate ancestor of various components of the adaptive immune system was one of the cell adhesion molecules (CAM) involved in the initial stage of embryonic organogenesis is the assumption that four β_2-microglobulin-like domains of N-CAM arose *in situ*. Were they borrowed from other molecules (even from immunoglobulins themselves) by the so-called domain exchange, the whole notion of CAM being the immediate ancestor of various components of the adaptive immune system becomes ridiculous. Fortunately, it looks as though these β_2-microglobulin-like domains of N-CAM indeed evolved *in situ*, for there is a noticeable similarity in construction of these β_2-microglobuin-like domains and other parts of N-CAM. Each of these β_2-microglobulin-like domains contains three absolutely invariant residues; 1) Cys in 12th position, 2) Trp in 24th position, and 3) Cys in 62nd position. These three invariant residues tend to be included in Thr-X,Thr-X dipeptidic repeats. This is illustrated at the top of Figure 2 on 3rd of the four successive β_2-microglobulin-like domains, for 3rd is the only complete domain, the other three sustaining deletions of three to six residues (Hemperly et al.,1986). Four successive β_2-microglobulin-like domains comprise but 40% of N-CAM polypeptide chain. The 362-residue long carboxyl terminal domain remaining within the cell is constructed of a simpler mode, thus, suggesting that this segment remained close to the original design of the entire CAM polypeptide chain. As also shown at the top of Figure 2, 699th to 728th residues of N-CAM is esentially made of Thr-X, Thr-X dipeptide repeats. Thus, it is conceivable that the entire coding sequence for the ancestral CAM was simple repeats of something like A C T C C A A, β_2-microglobulin-like domains too evolving from parts of it. Three consecutive copies of such a heptamer, 21 bases in the total length would have given the heptapeptidic periodicity to the original peptide chain as shown below:

THR PRO ASN SER LYS LEU GLN
A C T C C A A/A C T C C A A/A C T C C A A

Two base substitutions affecting the above noted periodicity unit would have produced three consecutive Thr-X dipeptides as shown below; two substituted bases are underlined:

THR PRO THR SER THR LEU GLN
A C T C C A A /<u>C</u> C T C C A <u>C</u>/A C T C C A A

At the top of Figure 2, A C T C portion of this hypothetical heptameric unit and its single base substituted deviants are solidly underlined.

Is there any validity to the above noted proposal as to the ultimate origin of CAM coding sequences. The first CAM must have come into being when the first multicellular eukaryote evolved from unicellular eukaryotes. Slime molds of the genus *Dictyostelium* indeed occupies a unique position of being an intermediate between unicellular and multicellular eukaryotes, for these organisms in nutrient rich environments live as unicellular

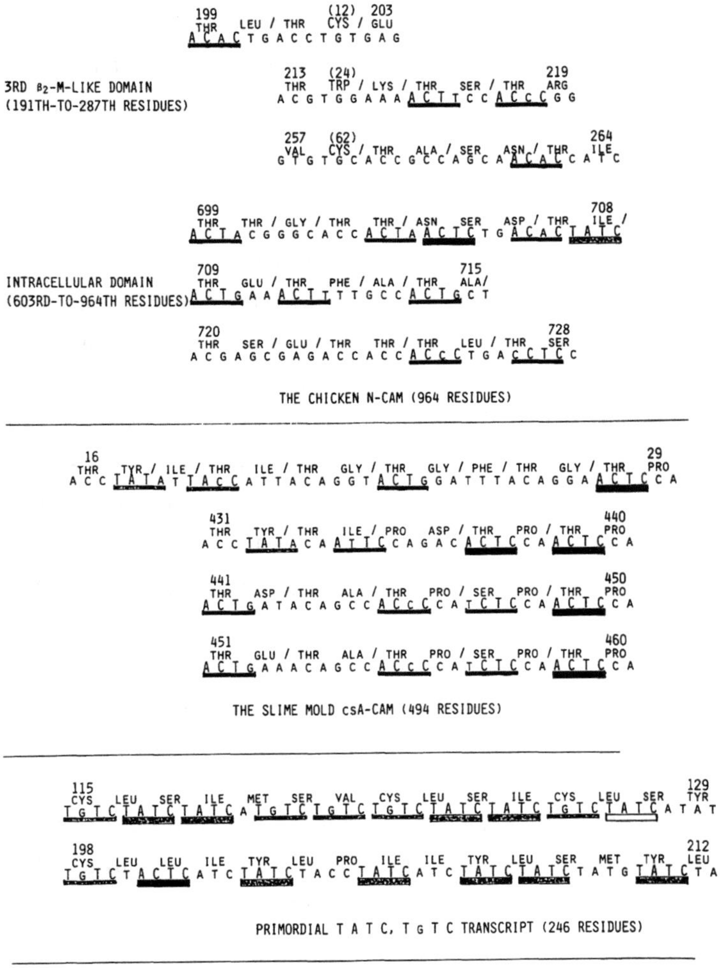

Figure 2

The indication of propinquity of descents between the Chicken N-CAM at the top (Hemperly et al.,1986), csACAM of the slime mold (*Dyctiostelium discoideum*) in the middle (Noegel et al.,1986) and the mouse transcript of primordial T A T C, T G T C repeats (Ohno and Epplen,1983). At the top, internal homology within N-CAM between β_2-microglobulin-like domains and the apparently more ancient intracellular domain is indicated. Within each β_2-microglobulin-like domain, three most invariant residues are a pair of cysteine for the intradomain disulfide bridge (12th and 62nd positions) and TRP at 24th position. As exemplified in 3rd β_2-microglobulin-like domain of the chicken N-CAM, THR-X,THR-X dipeptidic repeats invariably occur in vicinities of these three most invariant residues. THR-X,THR-X dipeptidic repeats are even more prominent feature of the intracellular domain. The principle tetramer A C T C and its single base substituted deviants are solidly underlined. One T A T C primordial tetramer is identified by a shadded bar. Although not identified, both T G T G and T G A C tetramers recurr twice each in six short coding segments of the chicken N-CAM shown at the top. Both T G T G and T G A C are single base substituted deviants of T G T C. In the middle, four coding segments of the slime mold csA-CAM which are essentially encoding THR-X,THR-X dipeptidic repeats are shown. A C T C and its single base substituted deviants are again identified by solid bars. The 30 base-long tandem repeats are noteworthy. 441st to 450th codon differs from 451st to 460th codon by a single base. At the bottom, a portion of the mouse primordial transcript which is mixed repeats of T A T C and its single base substituted deviant T G T C is shown as the ultimate ancestor of CAM coding sequences.

amoeboid creatures. When surroundings become unfavorable, however they begin to aggregate with each other to form the stalk and fruiting body, much in the manner of fungal species that include various mushrooms. This aggregation is induced by cyclic-AMP and mediated through csA CAM, and the 494-residue-long amino acid sequence of *Dictyostelium discoideum* csA CAM has recently been deduced from cDNA base sequence (Noegel et al.,1986). Indeed, it appears as though this primordial CAM has evolved from Thr-X,Thr-X dipeptidic repeats as shown in the middle of Figure 2. Particularly noteworthy is the coding segment encoding 431st to 460th residues, for it is made of three consecutive copies of the 30-base-long unit. It would be noted that 2nd and 3rd copies differ from each other only by a single base substitution, while 11 base substitutions separate 1st from 2nd. The already noted tetrameric unit A C T C and its single base substituted deviants are again very prominent in csA CAM coding sequence. However, it appears that this is a derived oligomeric unit and not the original repeating unit. The A T/ G C ratio of A C T C tetramer is 50/50. But csA coding sequence is quite unusual in that 62.6% of the sequence is A and T. The original repeating unit of the ultimate ancestor of CAM coding sequences had to contain considerably more A and T than G and C. Thus, we come to the tetrameric repeat coding sequence of the mouse which we previously reported as one of the few ultimate ancestors of all coding sequences (Ohno and Epplen,1983). This primordial coding sequence is mixed repeats of two tetrameric units; T A T C and its single base substituted deviant T G T C as shown at the bottom of Figure 2. The even representation of two tetramers give to the primordial coding sequence A T/G C ratio of 62.5/37.5. It would be recalled that this is the exact ratio found in csA CAM of the slime mold. Indeed, overall, T A T C, T G T C and their single base substituted deviants are as prominent as A C T C and its single base substituted deviants in csA CAM as well as N-CAM coding sequences. However, the latter gains prominance in segments dominated by Thr-X,Thr-X dipeptidic repeats preferentially shown at the top and middle of Figure 2. However, it would be noted that a pair of invariant CYS of each β_2-microglobulin-like domain of N-CAM is invariably encoded by a part of T G T G tetremer which is a single base substituted deviant of T G T C as shown at the top of Figure 2. This applies to invariant Cys in components of adaptive immune system as well. Thus, we have deduced the ultimate ancestor of various CAM genes engaged in cell-cell recognition of early embryonic organogenesis as well as genes for various components of the adaptive immune system to mixed repeats of two base tetramers T A T C and T G T C.

THE PRINCIPLE OF RECURRING UNITS IN CONSTRUCTION OF CODING SEQUENCES, LANGUAGES AND MUSICAL COMPOSITIONS

In our galaxy and others, stars have been formed and are still being formed by gravitational condensation of molecular clouds that contain large quantities of molecular hydrogen, water, ammmonia, carbon monoxide, methyl alcohol, hydrocyanic acid and others. When the earth was formed some 4.5 billion years ago, the primeval atmosphere surrounding it must have also contained these chemically reducing compounds noted above (Holye,1979;

Dyson, 1985). In the classical experiment of Miller in 1953, electric sparks passed through a mixture of methane, ammonia, molecular hydrogen and water yielded large fractions of amino acids; notable being alanine of a 2% yield. Oro in 1960, on the other hand, prepared a concentrated solution of ammonium cyanidein water. After a period, he found spontaneous converison of ammonium cyanide to adenine with 0.5% yield, (Miller and Orgel, 1974). Thus, it might be said that the yielding of various building materials of life was and is inherent in the composition of molecular clouds. What is life but a form that reproduces near exact replicas of itself. Thus, we owe our lives to the inherent complementarity that exists between the two purine-pyrimidine pairs of bases. Adenine pairs with uracil or thymine, while guanine forms hydrogen bonds with cytosine. Accordingly, when two complementary strand of double-stranded nucleic acids fall asunder, each can form its complementary strands. By this way, nucleic acids are inherently designed to perpetuate their base sequences. Inasmuch as the copying of the template, that is to say building of a new single stranded RNA complementary to the preexisted single stranded RNA is based upon the above noted inherent complementarity between A and U as well as G and C, this could have taken place in the prebiotic world, for if provided with a template as long as 60 to 100-base-long, A T P, G T P, U T P and C T P would align themselves in the proper 3'-5' linkage to form a complementary strand in the presence of Zn^{++} metal ion alone (Bridson and Orgel,1980). The major obstacle in the prebiotic world against spontaneous generation of the first cell on this earth, thus, was the formation of long enough templates directly from A T P, G T P, U T P and C T P, for even in the presence of imidazol and Zn^{++}, autopolymelization of nucleotide triphosphates yields only base hexamers to decamers. It follows then that unless these base oligomers were endowed with the inherent property for self elongation, long enough templates would not have come into being to start life on this earth. What if a given base octamer was repeats of the base tetramer such as T A T C already noted This octamer and its complementary strand formed after the first round of copying may have reannealed unequally first copy to the second copy after falling asunder as illustrated below:

T A T C T A T C
: : : :
A T A G A T A G

The hydrogen bonded paired portion would have served as a primer for the next round of copying (replication), and after this round, the octameric template would have elongated itself to the dodecameric template. Indeed, self elongation is inherent in repeats of base oligomers (prebiotic nucleic acids were RNA rather than DNA, thus, two T's of T A T C should have been substituted by U's, but for the sake of continuity, U A U C is shown as T A T C). This, then, is one of the many reasons for believing that the first set of coding sequences emerged at the very beginning of life on this earth were all repeats of base oligomers (Ohno and Epplen,1983).

Indeed, we have already seen that mixed repeats of T A T C and its single base substituted deviant T G T C appear to have served as the ultimate ancestor of one superfamily of genes; first various CAM's for general cell-cell recognition during the initial stage of organogenesis of all multicellular eukaryotes and through them, various components of the adaptive immune system unique to vertebrates. It would be noted that such tetrameric repeats resemble Julius Caesar's remark already cited in construction. *Vi'di* in the middle can be considered as T A T C, then *ve'ni* preceeding it becomes its two base substituted copy such as T G A C, while *vi'ci* following it becomes its single base substituted copy such as T G C C.

Such tetrameric repeats also resemble musical compositions of the Baroque period. As an example, the treble clef musical score of Prelude No. 1 for well-tempered clavichord by Johann Sebastian Bach (16856-1750) is shown in Figure 3. It would be noted that the initial part of this treble clef score in C major (the top 2 and 2/3rd lines of Figure 3) is essentially four note repeats; the second half of each 8/8th time signature segment being the exact copy of the first half. Each half of the time signature segment is comprised of two sets of the identical four notes; 4th note of the 1st set overlapping with 1st note of 2nd set. From the last one-third of the 3rd line of Figure 3 and downward, the theme now changes to three note repeats. This is because 1st note of each previous four note unit is now relegated to the base clef score. Such striking resemblance between Baroque musical compositions and primordial coding sequences that are repeats of base oligomers tempted us to devise one invariant rule by which treble clef scores of musical compositions and coding base sequences become interchangeable. After considering their respective molecular weights and complementarity, we have decided to assign a space and a line above it of the treble clef staff to each of the four bases in the ascending order of A G T C; C on the line of the previous scale occupying the classical middle C position (Ohno and Ohno, 1986). This assignment of bases to the treble clef staff afforded a needed freedom in transmutating coding base sequences to treble clef musical scores. This freedom is analogous to that accorded to coding sequences by the redundancy of

Figure 3

The treble clef score of an initial portion of J.S. Bach's Prelude No.1 from well-tempered clavichord is shown accompanied by a base sequence transcribed from it according to the pre-

Figure 4,Part I

Figure 4,Part II

Figure 4, Part III

viously devised invariant rule (Ohno and Ohno,1986). Initial tetrameric repeat portion should have encoded a polypeptide chain of tetrapeptidic periodicities, except for an unfortunate concentration of chain terminaters T A A's and T A G's at the extreme right of 2nd line. Subsequently, the treble clef score becomes trimeric repeats monotonously encoding homoserines occasionally interspersed by stretches of homoisoleucines and homoarginines.

base tetramer A G C A; last A of the first unit overlapping with 1st A of 2nd unit, and the same with regard to 3rd and 4th units. This (A G C A) X 4 recurrs as 4th segment. The 2nd segment, on the other hand, appears as four repeats of A T C A; A T C A being a single base substituted deviant of the previous A G C A. However, it would be noted that all four notes of the first segment unit changed a step each in the second segment unit; in musical notation from e g b d to d h c f. In our devised rule, however, only two of the four possible single step changes can be detected as base substitutions; from a position on the line to a space above as well as from a position in the space to a line below. Whereas two other single step changes, from a position on the line to a space below as well as from a position in the space to a line above, are perceived as synonymous.

In compliance with this rule, a portion of the primordial T A T C, T G T C repeats corresponding to 91st to 144th codon in its longest open reading frame (Ohno and Epplen,1983) has been transmutated to the musical score in A minor and 8/8th time signature as shown as part I of Figure 4. This is to be regarded as prelude, for as part II of Figure 4, the transmutation in C major and again in 8/8th time signature of 431st and 464th codons of the slime mold csA CAM coding sequence is shown. As shown in the middle of Figure 2, this portion of csA CAM coding sequence is comprised of three copies of the 30-base-long unit. The unit itself, however, apparently arose as repeats of shorter oligomers. One such tetramer, A C T C and its single base substituted deviants are identified by solid bars. This evolutionary trilogy ends in Part III of Figure 4 which celebrates the birth of original β_2-microglobulin-like domains in N-CAM -like cell adhesion proteins. The initial one-third of the coding sequence for 3rd β_2-microglobulin-like domain of the chicken N-CAM (Hemperly et al.,1986) has been transmutated to the treble clef musical score of Part III. Accordingly, Part III contains the first CYS for the invariably present intradomain disulfide bridge (in the middle of 2nd line of Figure 4, Part III) as well as the equally invariant TRP seen at the extreme right of 3rd line of Figure 4, Part III. Both CYS and TRP noted above are parts of THR-X, THR-X dipeptidic units. Even though the coding segment depicted in Part III is comprised of only 108 bases, there are still base oligomers recurring within. Tandem repeats of the pentamer G A T C A is seen at the extreme right and extreme left of 1st line, and that the base octamer C T T C C A T C encoding 210th to 213th PRO-SER-ILE (in the middle of 3rd line of Part III) is a single base deviant of C T T C C A c C encoding 217th to 219th THR-SER-THR seen straddling 3rd and 4th line of Part III. These recurring base oligomers still provide a melodious quality to aged coding sequences a billion or more years removed from their ultimately ancestral oligomeric repeats.

SUMMARY

Common denominators in all our cognitive processes are recurring elements. For example, the first step in deciphering ancient writings left on excavated tablets of a long lost civilization would be to identify the most frequently recurring set of symbols, for such a set likely represents the main subject with which those writing were concerned; be it a king of a particular dynasty or a taxable unit of lands. Similarly, our vision perceives patterns as a pattern only if a pattern is repeated, and a melody becomes a melody only when it is repeated. The same applies to all components of the adaptive immune system. All together they form a cognitive pattern because they were all derived from the ancestral β_2-microglobulin-like unit which probably arose in cell adhesion molecules (CAM); those plasma membrane proteins through homologous recognition contributed and are still contributing to the initial stage in organogenesis of all multicellular eukaryotes. This reliance on repetitions of our biological system appeared to have started at the very beginning of of coding sequences were likely to have been repeats of base oligomers. I have composed a musical trilogy to celebrate the birth of original β_2-microglobulin -like domains in CAM-like molecules. Part I represents the ultimately ancestral mixed repeats of T A T C and its single base substituted deviant T G T C, Part II depicts a portion of csA CAM coding sequence of the slime mold (a link between unicellular and multicellular eukaryotes) which encodes THR-X, THR-X dipeptidic repeats. Finally, Part III represents 3rd β_2-microglobulin-like domain of the chicken N-CAM. On one hand, this symbolizes the immediate ancestor of all the components of the adaptive immune system. On the other hand, it is linked to the past through recurring THR-X,THR-X dipeptidic repeats.

REFERENCES

1. Bridson, P.K, and Orgel, L.E. (1980) Catalysis of accurate poly (C) directed synthesis of 3'-5' linked oligoguanytes by Zn^{+2}. J. Mol. Biol. 144:567-577.
2. Dyson, F. (1985) Origins of life. Cambridge Univ. Press, Cambridge, London.
3. Hemperly, J.J., Murray, B.A., Edelman, G.M. and Cunningham, B.A. (1986) Sequence of a cDNA clone encoding the polysialic acid-rich and cytoplasmic domains of the neural cell adhesion molecule N-CAM. Proc. Natl. Acad. Sci.USA 83:3037-3041.
4. Hoyle, F. (1979) Ten faces of the universe. Freeman Press, London.
5. Miller, S.L. and Orgel, L.E. (1974) The origin of life on the earth. Prentice-Hall, New York.
6. Noegel, A., Gerisch, G., Stadler, J. and Westphal, M. (1986) Complete sequence and transcript regulation of a cell adhesion protein from aggregating *Dictyostelium* cells. Embo J. 5:14731476.
7. Ohno, S. and Epplen, J. (1983) The primitive code and repeats of base oligomers as the primordial protein-encoding sequence. Proc. Natl. Acad. Sci. USA 80:3391-3395.
8. Ohno, S. and Ohno, M. (1986) The all pervasive principle of repetitious recurrence governs not only coding sequence construction but also human endeavor in musical composition. Immunogenetics 24:71-78.

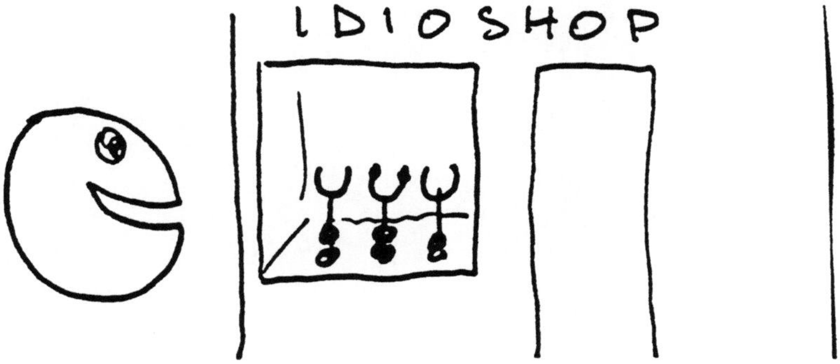
IDIOSHOP

STRUCTURAL COUPLING AND THE ORIGIN OF MEANING IN A SIMPLE CELLULAR AUTOMATON

Francisco J. Varela

CREA
Ecole Polytechnique
1, rue Descartes,
Paris 75005

1.Introduction: Classification vs. Generation

In this paper I wish to address what I consider one of the central conceptual issues underlying the Immunosemiotics meeting: the origin in meaning in cellular communication. Semioticists have done immunology a great service by framing the role of molecular interactions as a system of signs and their significations. This is hardly what immunologists normally do, involved as they are with the nitty-gritty of molecular mechanisms, far too close to visualize the entire forest.

In my view, however, current semiotics is not a satisfactory framework to advance understanding of cellular communication because so far it has only classified signs and their systems [1]. There is nothing by the way of a generative mechanism as to how a sign becomes one: there is no explicit mechanism for the origin of meaning. It is somewhat like reading a taxonomy text before natural selection was proposed as a mechanism for the origin of diversity. This contrast --taxonomic (or structuralist) vs. generative-- was, in my opinion, at the center of many discussions during the meeting.

Accordingly, my intention in this paper is to address this issue directly through the use of a simple example. I hope that this will do more to illustrate what I have in mind that a dry theoretical presentation. Some conceptual background is nevertheless necessary , and that is the topic of the next section.

[1] I am using as a reference here U.Eco, *A Theory of Semiotics,* Indiana U.Press, 1978.

NATO ASI Series, Vol. H23
The Semiotics of Cellular Communication in the Immune System
Edited by E. E. Sercarz et al.

2. Operational Closure and Structural Coupling.

Meaning can only arise for those systems which assert their own identity vis-a-vis their environment, that is, for systems with a degree of autonomy. This class of autonomous systems stands out in contradistinction to those systems which are defined through input/output relations. For these latter class of systems, the way in which they relate to their environment is not an issue: they are specified through input/output conditions which completely clarify how their environmental coupling occurs. The familiar digital computer is the clearest example of this class of systems: the meaning of a given keyboard sequence is always assigned by the designer.

In contrast, cells and organisms are far from being in the same category. Under very restricted circumstances we can speak as if we could specify the operation of a cell or an organism through input/output relationships. In general, however, a living system brings forth its own world of relevance, and is not given in advance. The meaning of this or that interaction is not given by an outside designer, but is the result of the organization of the system itself and its history. This point is intuitively quite evident to any working biologist, but more often than not, eschewed because of its seeming difficulty.

The question is, then, how to account for the emergence of meaning from a milieu of interactions for a system which cannot be specified through a list of input/outputs. The answer I have been developing for some years is basically this: follow the consequences of characterizing the system, not as input/output, but as having operational closure, that is, as defined through a network of processes of some kind. Once a system is adequately defined in this fashion, its self-organizing qualities will become apparent. In particular, a history of coupling with an environment will lay down what is and what is not relevant for the system: the history of coupling will inevitable bring forth a world of signification[2].

---

[2] I'm painfully aware that these key notions --operational closure and structural coupling --remain quite vague here. I only intend to evoke some of their signification and to use them through the example. For a full discussion see: F.Varela, *Principles of Biological Autonomy*, Elsevier/North Holland, New York, 1979, and for an introductory account see H.Maturana and F.Varela, *The Tree of Knowledge: The biological roots of understanding*, New Science Library, Boston, 1987.

My experience in trying to convey these ideas is that although many people accept with some ease the notion of operational closure (self-organization in a network -like system), the complementary notion of the origin of meaning through structural coupling remains veiled in a cloud of mystery. The reason for this is not intrinsic complexity, but simple lack of occasion to be exposed to such an analysis. This is why I wish to provide an example that is so transparent that it can serve as a paradigm or exemplar.

3. The System

The example I wish to present is based on very simple cellular-like automata which receive inputs from two immediate neighbors, and communicates its internal state back to them. Each cellular automaton can only be in two alternative states (i.e. 0 and 1, active or inactive). Assume that the rule governing the change in each automata is simply a boolean function of the two arguments it receives from its neighbours at each discrete moment of time. Typical boolean function are, for instance, the well-known logical operation 'And' or 'Exclusive or'. We can associate such a boolean function to each one the two states in which the cellular automaton is in; this says that the way the cellular automaton will respond to its surrounding neighbours will change depending on its own state. Thus, the functioning of each automaton is completely specified by a pair of boolean functions. In our figures this is expressed as an 8-digit binary number if we express each boolean function in terms of their 4-digit transition tables, which indicates the output of the automaton as a function the four possible combination of its two inputs .

The closure of the system is now introduced simply by connecting it as a circular array, so that there no inputs and outputs from the entire system ring, but only internal actions. For the purpose of display, however, it is easier to cut this ring open and to present it linearly, with the cells in the 1 state indicated by a black square, and the opposite state indicated by a blank space. Accordingly, in the display shown here, cellular position runs from left to right (modulo the ring's length) (Fig.1).

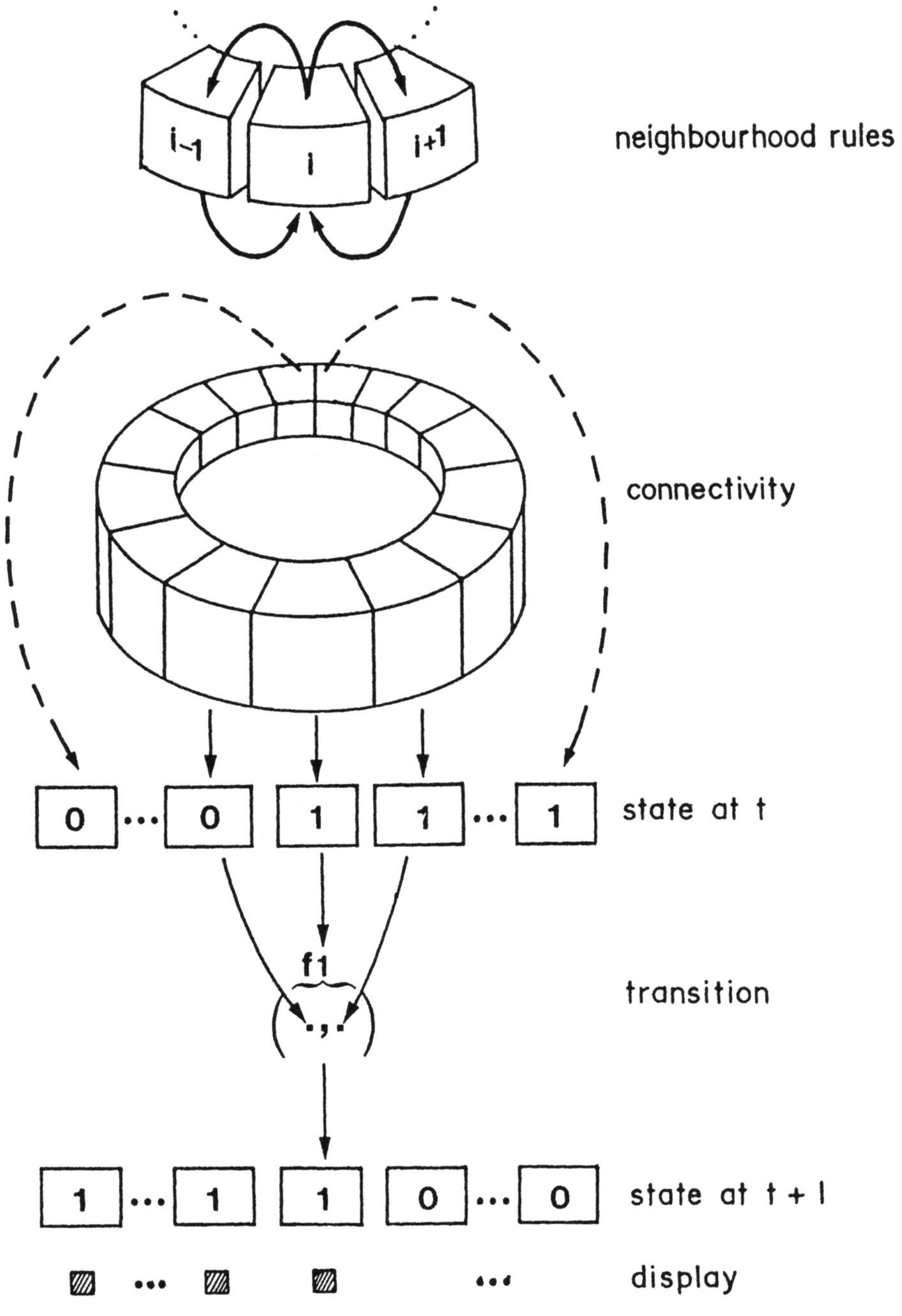

Figure 1

Schematics for the definition of the cellular automata studied here (see text).

This ring of cellular automata acquires a temporal dynamics by starting it at some random state and letting each cell reach an updated state at each discrete moment of time in a synchronous fashion (i.e. all of them together). In the display we represent the initial instant at the top-most row, and successive instants of time going downwards. Thus the successive state of the same cell can be read as a column, and the simultaneous state of all cells can be read as a row.

In the cases we study here all cells in the ring are defined by the same pair of boolean functions, and it appears inscribed on top of the display. Further, we consider only symmetrical boolean functions (i.e. with identical entries in the diagonal of their transition matrix, so that strictly speaking we need only six digits to define the dynamics of the ring). In all the simulations presented here the ring was composed of 80 cells, and its initial starting state was chosen at random.

4. The dynamics of the closure: self-organizing qualities

These rings of boolean cellular automata have very interesting and surprising self-organizing capacities. An extensive discussion of these capacities has been conducted recently by S.Wolfram [3]. It is not my intention to recapitulate his work here. Suffice it to say that the closure dynamics of these rings fall into four major classes illustrated in Fig.2. A first class exhibits a simple dynamics dominated by a single attractor, leading all cells to become homogenously active or inactive. For a second more interesting class of rings, the rules give rise to spatial periodicities, that is, some cells remain active while other do not. For a third class the rules gives rise to spatio-temporal cycles of length two or longer. Finally, for a few rules, the dynamics seem to give rise to chaotic regimes, where one does not detect any regularities in space or time.

We shall not dwell on the interesting diversity of the internally generated dynamics of these boolean rings. Our interest is to move beyond their closure to examine what happens when they enter into a history of coupling.

---

[3] See in paricular: S.Wolfram, Statistical mechanics of cellular automata, *Revs.Modern Physics* **55**: 601-644, 1983, and: Cellular automata as models of complexity, *Nature* **311**:419, 1984.

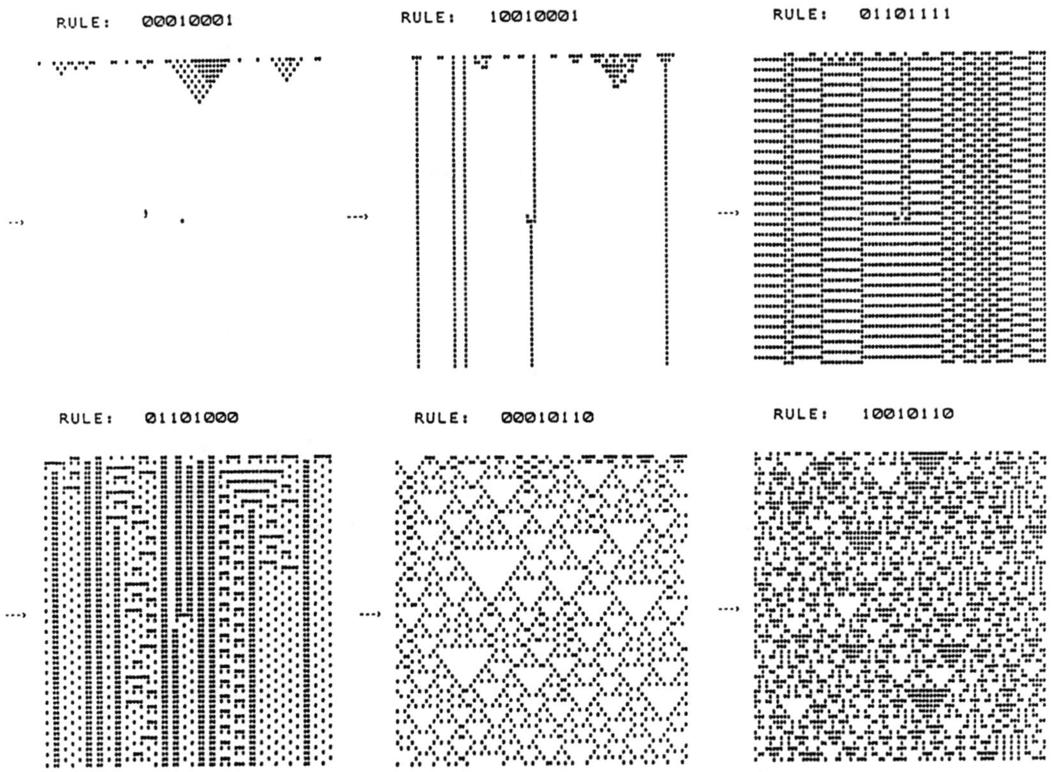

Figure 2

Representative self-organizing dynamics for different choices of defining rules (the 8-digit binary number on top). Initial states in all cases is randomly assigned. The differences in the global behavior is immediately visible in the pattern of the display. Arrow to the left indicates the arrival of a perturbation following the mode of coupling depicted in Fig.3.

5. Histories of structural coupling.

Of the many forms in which we could endow our rings with a coupling with an environment, I have chosen the following. Imagine that we dump the ring in a milieu of random 0's and 1's, much like a cell is plunged in a chemical milieu. Imagine further that the encounter of a cell with one of these two alternatives (0's and 1's) leads to the state of the cell being replaced by the perturbation that it encountered (Fig.3). For the sake of brevity, let us call Bittorio this particular ring of cellular automata <u>with</u> this form of structural coupling with the chosen milieu.

In Fig.2 the arrow to the left indicates the moment where one perturbation reaches one particular cell at one particular instant. The dynamics that follows indicates the way Bittorio compensates this perturbation, i.e. the ensuing change (or lack of it). As it is apparent, if Bittorio's rule belongs to the first or fourth class (i.e. a simple or a chaotic attractor) the consequence of the perturbation is simply invisible: Bittorio either goes back to its previous homogenous state, or it remains in a random-like state.

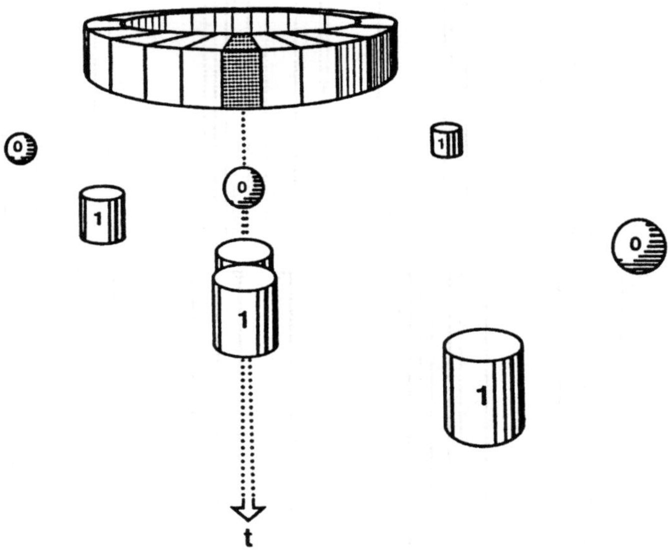

<u>Figure 3</u>

Schematics for the mode of coupling defined for the cellular automata of Fig.1. A chance encounter with a 1 or 0 changes the internal state of the perturbed cell for that of the perturbation encountered. For simplicity, perturbations at only one location are displayed here.

It follows that only the second and third class of rules can provide us with a dynamics capable of producing interesting consequences for the kind of structural coupling we have chosen for Bittorio. As also shown in Fig.2, for these kinds of Bittorios, a single perturbation induces a change from one to another spatio-temporal configuration, both of them stable and distinguishable.

The base of Bittorio of rule 1001000, illustrated in Fig.4, is worth commenting in more detail. As can be seen , the encounter with just one perturbation changes the spatial periodicities from one to another stable configuration. However a second perturbation at the same cell undoes the previous change. Hence, any odd sequence of perturbations at the same locus will lead to a change in the state configuration for Bittorio, while any even sequence of perturbations will be invisible, since it leaves Bittorio's global state unchanged. Thus, of all the innumerable sequences of possible perturbations, this Bittorio picks up or singles out from the milieu a very specific subset: finite odd sequences; only they induce a repeatable change in

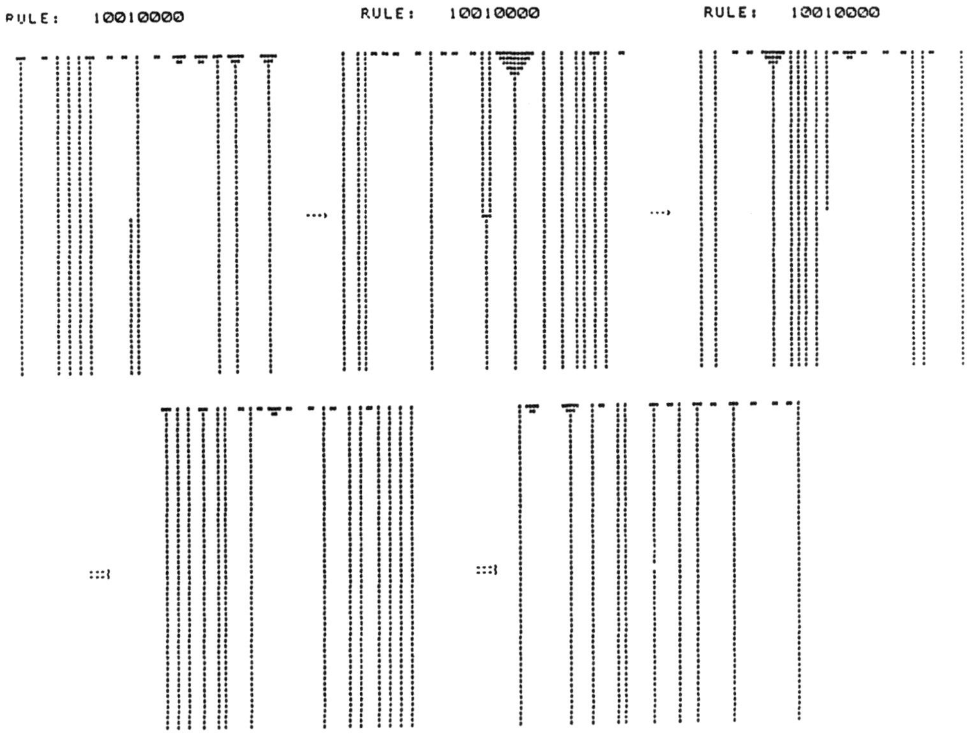

Figure 4

An instance of discrimination of odd sequences of perturbation. Automata with the same rule but different initial configurations are displayed. Top row: encountering one perturbation at arrow. Bottom row: encountering two successive perturbations.

configuration. Stated in other words, given its rule, and given its form of structural coupling, this Bittorio become an 'odd sequence recognizer'

Another example of this emergent signification is shown in Fig.5 for Bittorio of rule 01101110. A sequence of two perturbations is the only trigger capable of leading to a change in the state configuration of Bittorio. This is readily seen in Fig.5 where I have superimposed several encounters at different cellular loci to facilitate comparison. Everything other than double perturbations in one location leave this Bittorio unchanged.

I have explored what happens to these ring automata with simultaneous perturbations and more complex forms of structural coupling. They reveal very many rich and interesting behaviors. I will not enter into the detail of these observations for the purpose of illustration: our simple Bittorio is enough.

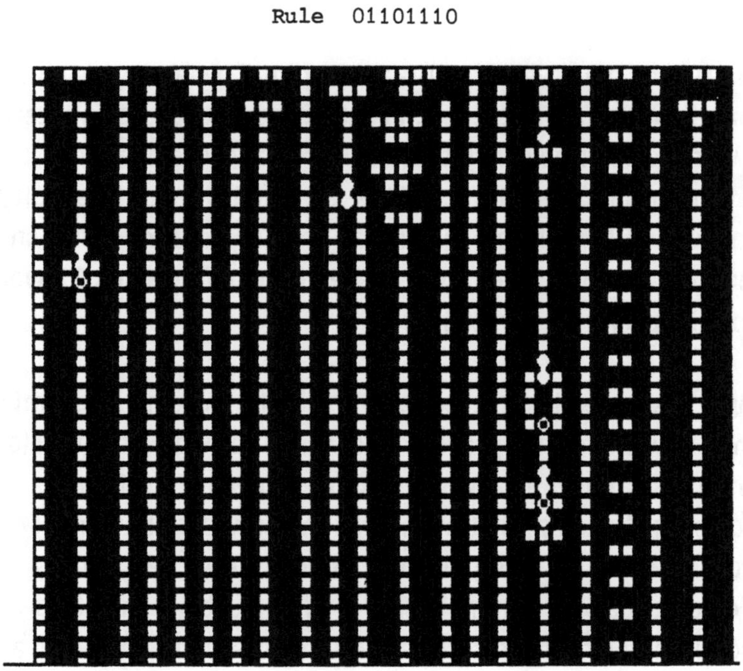

Figure 5

An instance of discrimination of pairs of perturbations. Here, in contrast to Fig.4, several sequences of perturbations are induced at different locations on the same automata, since each one only has local effects.

6. Bringing forth meaning.

In the two specific cases described above, we have not provided Bittorio with a program to distinguish 'odd sequences' or 'two successive perturbations'. What we have done is to define a form of closure on the one hand (its internal dynamical alternatives), and the way in which this system will couple with a given milieu on the other hand (the contingence of its encounters). The result is that the history of this coupling brings forth or selects from a random world what is relevant for the system itself. The system recognizes out of its own autonomy what is significant and how it is so.

I have used the words significance, relevance, and meaning advisedly. Meaning involves necessarily a form of interpretation in an encounter. My claim is that Bittorio performs precisely this form of a minimal interpretation in its encounters, thus pointing to us, as observers, what does and what does not lead to changes in its internal dynamics.

This is similar to the study of cellular or animal behavior. We must see what the animal singles out as relevant in reference to changes in internal dynamics; its world cannot be constituted as a set of predefined inputs independently of its particular history. This is the interpretive capacity upon which the creativity of life hinges. In general, then, my claim is that once closure and coupling are described for a system, the emergence of a world of relevance becomes apparent to an observer.

The example presented here is extremely simple, in some sense minimal, so that we can follow the entire process in detail. That was its intention; it should not be read as a model of any specific phenomena of cellular recognition. However it seems that even with the very simple form of operational closure and of coupling given to Bittorio if we can already recognize the emergence of a minimal form of signification, then this will be all the more so for living cells or cellular networks such as the brain and the immune system.

Notice, however, that what I have proposed is a mechanism and not a recipe. To investigate how it is actually embodied in each situation, one needs to pursue the details of that situation. In particular, one must have at least some idea of:

1. the operational closure of the system and its dynamical landscape, and
2. the fundamental dimension of the structural coupling with the milieu.

Both these points are central for the network perspective in immunology [4]. In fact, since its inception this perspective has emphasize the self-organizing dynamics that must constantly arise in such a network system, in contrast to the local character of the clonal selection dynamics. Furthermore, the network perspective has always raised the issue of the dependency between internal dynamics and antigenicity, in contrast to the antigen-centered clonal perspective. In a word, the traditional immunological views are closer to input/output descriptions, and it is not surprising that the emergence of molecular significations remains difficult to handle within that framework. In contrast, a network perspective is already a step in the right direction to understand immune phenomena as a system of signs and significations produced by the immune system itself.

ACKNOWLEDGEMENTS

My thanks to Alfonso Gómez, Patricio Huerta and Marcelo Miranda for participating at various stages in the computer presentation of these ideas. My interest in immunology is due to the patient and friendly influence of Nelson Vaz and Antonio Coutinho. I hope it will continue.

FV is the holder of the Fondation de France Chair of Cognitive Science and Epistemology at Ecole Polytechnique. Thanks also to the Prince Trust Fund for financial support.

[4]The classical paper by N.Jerne, Towards a network theory of the immune system, *Ann.Immunol. (Paris)* **125C**: 373-395, 1974 emphasized the "eigen-behaviors" (or self-generated) states of the immune systems. The idea is also discussed in N.Vaz and F.Varela, Self and non-sense: An organism-centered approach to immunology, *Med.Hypotheses* **4**: 231-257, 1978 together with the notion of a cognitive domain. See also A.Coutinho, L.Forni, D.Holmberg, F.Ivars, and N.Vaz, From an antigen-centered clonal perspective of immune responses to an organism-centered network perspective of autonomous activity in a self-referential system, *Immunol.Revs.* **79**: 151-168, 1984.

The Control of T Lymphocyte Activity May Involve Elements of Semiosis

Klaus Eichmann
Max-Planck-Institut für Immunbiologie, Stübeweg 51, D-7800 Freiburg, FRG

T lymphocytes are the major functional cell type in cell-mediated immunity. While most immune defense events involve combinations of antibody and cell-mediated mechanisms, several infectious diseases are faught against by T lymphocytes only, without the participation of antibodies. The study of such diseases, notably leprosy, opens a particularly clear view into the properties of the T cell compartment, into the questions that have to be asked in order to come to an understanding of its ways of function, and why we need cell-mediated immunity in the first place.

Leprosy is caused by Mycobacterium leprae, a socalled facultative intracellular bacterium. This organism has developed at least two strategies to evade attack by any sort of soluble mediator such as antibodies. Firstly, the bacterium has a thick layer of lipid-rich material around its surface, making it relatively resistant to the lytic activity of antibody and complement. Secondly, the organism has developed the capacity to live inside macrophages, a cell type whose ordinary function in the body is to incorporate particles and to digest them. M. leprae are incorporated but survive and multiply in macrophages. Since antibodies are ineffective against intracellular antigens, this localisation is a most efficient way of avoiding antibody-mediated immunity. As a consequence, in order to survive M. leprae infections, the host needs a mechanism of immunity that is able to identify and destroy M.leprae infected macrophages. T lymphoytes have the capacity to distinguish infected from normal macrophages because they <u>recognize</u> as signs of abnormality small bacterial fragments on the macrophage surface. The ability of T lymphocytes to recognise abnormal cell surfaces is associated with a highly specialized recognition apparatus allowing them to distinguish soluble from cell associated antigens. Accordingly, T cells have the ability to recognize a bacterial fragment on a macrophage surface but ignore the same bacterial fragment in free circulation. Thus, the first basic question about the T lymphocyte therefore is that of the nature of its recognition apparatus, enabling it to recognize and to become activated by foreign components, but only if these are on the surface of other cells in the same body.

Presumably in order to survive in evolution, M. leprae needed and has developed a second level of evading even the mechanisms of cell-mediated immunity, but this second level is fully successful only in a few individuals. Leprosy occurs in a bipolar spectrum of clinical courses having two extreme ends: Socalled tuberculoid leprosy is

NATO ASI Series, Vol. H23
The Semiotics of Cellular Communication in the Immune System
Edited by E. E. Sercarz et al.

characterized by localized inflammation indicating vigorous immune defense, finally resulting in complete elimination of the infection and in recovery. On the other extreme end stands socalled lepomatous leprosy, in which there are no signs of immune reactions whatsoever and in which the patients finally carry incredible loads of bacteria in all parts of their body, eventually dying of the disease. This bipolarity gives rise to the second basic question in cell mediated immunity: How does the same microorganism induce in certain individuals of a species very effective T cell activation and, in others, no activation or even suppression of T cell immunity?

Some of the basic facts pertaining to these two questions are known: T cells possess on their membrane a type of glycoprotein molecule which, similar to antibody, is genetically put together from a multitude of different gene fragments thus occuring in a very large number of structural variants. Accordingly, this molecule, called the T cell receptor (TCR), can combine specifically with an equally large number of different structures. The force to combine with foreign structures only on the surfaces of other cells in the same body comes from the necessity of the T cell receptor to combine with complexes formed by the foreign structure with another type of cell surface glycoprotein, encoded by the major histocompatibility complex (MHC). The structural and functional basis of this complex formation and of the restriction of the T cell receptor to recognize such complexes are unknown but in the very center of immunological research today.

In all species studied so far, there exist two major types of MHC molecules (called class I and class II) and two major corresponding types of T cells, distinguished by a third type of surface glycoprotein. One T cell type carries a glycoprotein termed CD8 and its TCR combines with complexes of foreign structures with MHC class I molecules on other cells. The other T cell type carries a different glycoprotein, formed CD4, and its TCR combines with complexes of foreign structures with MHC class II molecules. Each of the two major T cell types may be composed of several subtypes. Although none of the T cell functions known to date are exclusively confined to one of the two major T cell types, it is a reasonable simplification to say that CD8 bearing, class I restricted cells are primarily concerned with the destruction of altered cells in the body (killer function), whereas CD4 bearing, class II restricted cells, induce other cells of the body to enhanced functional activity (helper/induced function). In addition there exist a T cell function termed "suppression" which is very poorly understood but appears, in the view of many investigators, as the result of the concerted action of a number of distinct T cell types. There is no agreement among immunologists which T cell types and which molecular mechanisms account for suppression. Nevertheless, as outlined above for the example of leprosy, suppression is a pivotal activity of the immune system, standing against successfull activation of immune defense mechanisms, possibly deciding between succumbing to or surviving an infection.

Experimentally, many investigators have demonstrated that T cell activity can be influenced by antibodies to various molecules on their surface. I describe here one set of experiments which to my opinion sheds a new light on the unsolved questions outlined above by suggesting a comparatively simple set of molecular interactions on T cell surfaces to be responsible for their state of activity. Using antibodies to TCR and to CD8 and CD4, Emmrich et al. (1) have shown that activation and inactivation of the T cell depends on the position of CD8 and CD4 relative to TCR. If CD8 cells are exposed to a mixture of anti-TCR and anti-CD8 antibodies such that crosslinking is induced between TCR and CD8 molecules, CD8 cells begin to express growth factor receptors, divide in the presence of growth factors, and become active cytotoxic cells. Analogous effects are induced in CD4 cells using antibodies to TCR and to CD4. If the experiments are performed such that no crosslinking between TCR and CD8/CD4 is induced, no activating effect is seen. Moreover, if each of the two types of molecules is independently fixed, thus preventing heterologous crosslinking, the T cell cannot be activated, i.e. is suppressed. The term "crosslinking" as used here implies that two or more molecules are fixed to each other, at a small distance, resulting in their immobilization relative to each other. These observations mean that a T cell reacts specifically to different modes of positioning certain molecules on its surface. If all molecules move independent of each other, the cell is in a neutral state, neither active nor suppressed. If the TCR becomes fixed to CD8/CD4, activation is induced. In contrast, if the TCR and CD8/CD4 become fixed by separate ligands, a state of suppression is generated.

These cell surface events may indeed resemble certain elements of semiosis. The same cell surface components appear to be able to form distinct patterns on the cell surface which are discriminated by the internal machinery of the cell resulting in distinct functional behaviour. There appears to be an element of choice in the response depending on the context in which the cells finds certain functionally crucial molecules on its membrane.

The analogy may go even further if one considers the putative physiological cell interactions which the described experiments may resemble. It is obvious that MHC (class I and class II) molecules represent in vivo ligands for the TCR. It is not proven but suspected with good reason that MHC class I and class II molecules also represent ligands for CD8 and CD4, respectively. Moreover, many investigators may accept that specific receptor molecules, TCR and antibodies, frequently bind to each others idiotopes in the same immune system. With these receptor ligand pairings the postulated activating and suppressive situations are most readily constructed: Simultaneous binding of the TCR and of CD8 (CD4) to the MHC class I (class II) molecules on another cell induces crosslinking of TCR to CD8 (CD4) and thus will result in T cell activation. As shown in Fig. 1, there are two versions of this mechanism, depending on the mono- or multivalency of CD8/CD4. With monovalent CD8/CD4, tertiary complexes between TCR, MHC, and CD8/CD4 would result. With multivalent CD8/4, larger

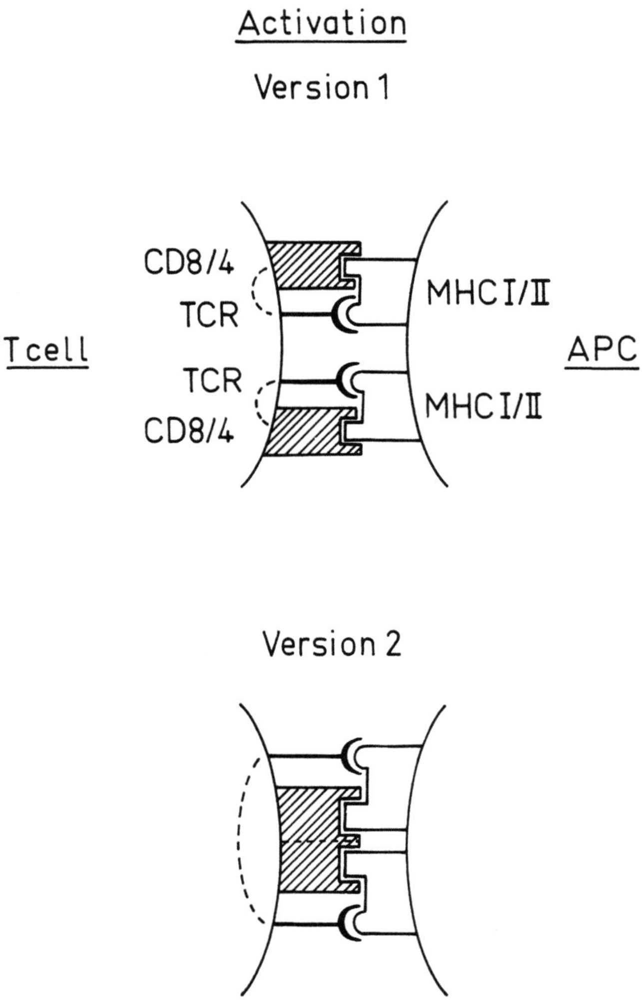

Fig. 1

Suppression

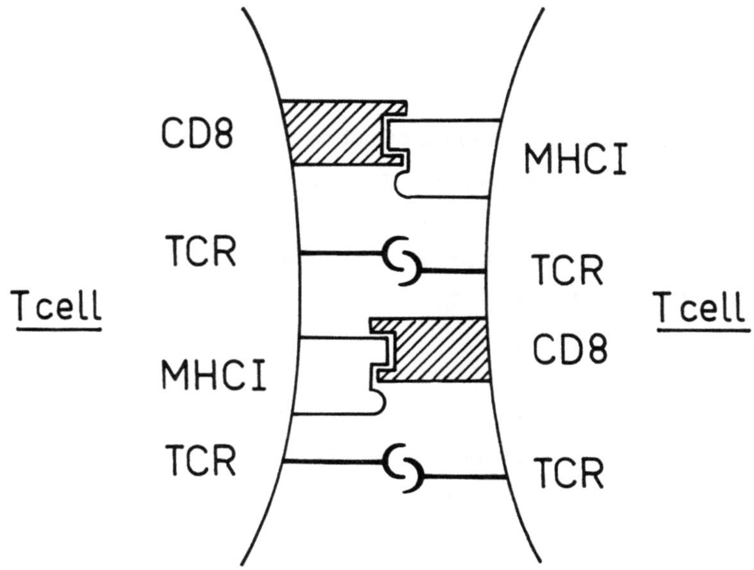

Fig. 2

complexes would be generated. Either type of interaction is expected to occur when a T cell interacts with an antigen presenting cell. In contrast, when a T cell interacts with another lymphocyte, its receptor may become engaged to the receptor and its CD8(CD4) may become engaged to the MHC class I (II) molecule of the other cell. As shown in Fig. 2, this generates a situation of separate fixation of TCR and CD8/CD4 and is expected to result in suppression. Again, the analogies to semiosis are obvious: Messages are conveyed between different members of the immune system by the way they engage each others surface components. There seems to be a choice between at least two functional alternatives.

The modes of cell interaction postulated for activation and suppression offer a simple answer for one of the two basic questions on cell-mediated immunity outlined above. The force to concentrate on cell surfaces comes from the necessity of crosslinking the TCR to CD8/CD4, which functions only if both bind, at the same time, to an MHC molecule on another cell. Moreover, this requirement for T cell activation would result in the observed strong selection to only activate CD8 positive T cells with MHC class I specific receptors and CD4 positive T cells with MHC class II specific receptors. Such a selection would result in the observed strict associations between CD8/CD4 and MHC class I/class II specificity even if there was no prior selection of the repertoire. However, it is also possible that similar crosslinking requirements exist during T cell repertoire selection and are instrumental in the generation of the two distinct repertoires in ontogeny.

The second basic question, to respond or not to respond, is not readily brought into the context of the present model of activation and inactivation. I feel, however, that models for suppression that are based on the assumption of different modes of interaction between a limited number of cell types are more likely to be true than models based on a multitude of cell types, each specialized for a narrow functional aspect.

1. Emmrich, F., U. Strittmatter, and K. Eichmann, Synergism in the activation of human CD8 T cells by crosslinking the T cell receptor complex with the CD8 differentiation antigen. Proc. Nat. Acad. Sci. USA 83, 8298, 1986

DIGITAL PROCESSING AND IMMUNO-REGULATORY CIRCUITS

Leonore A. Herzenberg
Department of Genetics, Stanford University, Stanford CA 94305

Although I rather doubt that semiotics per se has a lot to contribute to immunoregulatory notation and concepts, there is little question that this area requires some innovative structuring to provide a more facile framework for conceiving and discussing ideas. I have no clear-cut view of how this is to be accomplished; however, since any attempts in this direction must begin with the development of a catalog of data and concepts upon which overall theory can be based, I have tried to identify several key aspects of my own work that seem important to these deliberations.

Some time ago we introduced an immunoregulatory circuit model that offered a new approach to thinking about antibody response regulation. (1) In essence, this model proposed the existence of bistable regulatory circuits that individually control the production of particular antibody molecules or sets of molecules expressing related idiotypes. These cell interaction circuits can be thought of as analogous to the electronic binary ("flip-flop") circuits that are central to modern digital computers. They can be actively switched into either an "on" or an "off" position by auxiliary cell interactions (helper or suppressor) and will maintain that position despite minor perturbations in the immunologic environment. Nevertheless, they remain capable of being switched to the alternate position when a sufficiently strong signal is encountered.

The cells and signalling interactions that we proposed for these circuits were chosen to exemplify the properties of the circuit and were not intended as more than a rough approximation of reality. The principle, however, of a set of central regulatory circuits that acquire an initial configuration as a result of immunization (or tolerization) conditions and then tend to maintain that configuration unless immunoregulatory conditions change drastically has proven extremely useful in our subsequent work on epitope specific regulation. In essence, viewing these circuits as controlling memory B cell expression provides an explanation for the otherwise perplexing differences in antibody responses that occur when the order of immunizations with carrier molecules and hapten carrier conjugates varies. (2)

Results from a large series of studies begun with immunizations using standard carrier and hapten-carrier conjugates (3) and extending now to vaccine production studies with synthetic peptides conjugated to carrier proteins (4) have verified the existence of some kind of bistable immunoregulatory mechanism functioning in the normal immune system. This mechanism is unlikely to be mediated

NATO ASI Series, Vol. H23
The Semiotics of Cellular Communication in the Immune System
Edited by E. E. Sercarz et al.

by the model circuits in the precise form that we proposed; however, the principle embodied in these circuits and perhaps even some of the cell interactions that were defined constitute an important base for future theoretical considerations.

Much of the thinking that went into the development and utilization of the circuits model that we proposed was conditioned by a rudimentary knowledge of basic electronic circuitry and computer programming techniques. The living organism is clearly not a computer and, although it owes much to the transmission of electrical signals, is clearly not definable simply in terms of flip-flops and other electrical gadgets. Nevertheless, we have found that much can be gained by searching for ways in which the behaviour of biological mechanisms can be understood through analogies with processes utilized in computer hardware and software systems.

Unfortunately, contemporary biologists tend to be poorly trained for this approach. By and large, our language and thought processes are less analytical and more phenomenologically oriented than those used in mathematics and engineering. This perhaps is the source of our creativity, but it also limits our ability to think and speak about our work. Therefore, I wonder whether the next conference in this series might be well spent considering the contribution that application of some of the concepts used in digital electronics and artificial intelligence would make in clarifying current thinking in immunology. Perhaps we could develop a synthesis that would expand our vocabulary without overly regimenting our minds.

References

1. Herzenberg, Leonore A., Samuel J. Black and Leonard A. Herzenberg. 1980. Regulatory Circuits and Antibody Responses. Eur. J. Immunol. 10: 1-11.
2. Herzenberg, Leonore A., T. Tokuhisa, K. Hayakawa. 1983. Epitope-specific regulation. Annual Reviews of Immunology, Vol. 1: 609-632.
3. Herzenberg, Leonore A., Takeshi Tokuhisa and Leonard A. Herzenberg. 1980. Carrier-priming leads to hapten-specific suppression. Nature 285: 664-666.
4. Schutze, M-P., C. Leclerc, F.R. Vogel and L. Chedid. 1986. Epitopic suppression in synthetic vaccine models: Analysis of the effector mechanisms. Cell. Immunol. 104: 79-90.

OBSERVER, IMMUNE SYSTEM AND THEIR RESPECTIVE OBJECTS (ABOUT SELF - NON-SELF DISCRIMINATION)

F. Jacquemart and A. Coutinho

Laboratory of Immunobiology
Pasteur Institute
25, rue du Docteur Roux
75724 Paris Cédex 15
France

The notion of self is perhaps one of the oldest in human history. This is precisely what makes it difficult to analyse. It is hardly possible to set up any operational procedure allowing for the qualification of an object as self or not self without referring to the very notion of self in that procedure.

In scientific discourse, it is customary to begin with something which apparently has some operational meaning. Thus, we will first use the notion of "self" as it is generally used without directly questioning this notion, so that we can first focus our attention on the context of that usage.

In Ehrlich's thesis, summarized by the "horror autotoxicus" (1), the lymphocytes able to produce antibodies against **"self elements"** are eliminated. In this framework, non autotoxicity obvioulsy requires toxicity towards autoreactive cells which are components of the "self". The notion of "horror autotoxicus" therefore contains a paradox. Of course, this paradox is only apparent, and the mistake is evident. However, we believe that it is the very evidence of this contradiction that makes it a good illustration of what may be genrally omitted in other situations. We cannot formulate any proposition concerning living systems without considering levels of organization and levels of description (this also means that something which concerns any behavior at the elemental level does not necessarily have an explicative value at the systemic level). In Ehrlich's thesis, lack of toxicity towards the individual requires the destruction of some components of the individual.

Without discussing the later modifications introduced into Ehrlich's thesis, we can apply directly the same argument in the case of the idiotypic network: even if we admit the non toxicity of the immune system to the body (and thus for the immune SYSTEM), this will tell us nothing about the behavior of elements between themselves. This is even more so if, instead of

NATO ASI Series, Vol. H23
The Semiotics of Cellular Communication in the Immune System
Edited by E. E. Sercarz et al.

toxicity we use the aptitude of an immunological element to bind such or such "self" component. In fact, if, in immunology, toxicity implies binding, the reverse is obviously not true. It is surprising that these two notions are so frequently presented or used as if they were complementary. It is not paradoxical to state at the same time that there exists a self - non-self discrimination by the immune SYSTEM on one hand, and that, on the other hand, this phenomenon implies binding, or even toxicity of elements towards other elements of the system. The principle here is the same as in the Ehrlich case: we must distinguish between levels of organization, and this limits the scope of our general propositions. The immune response, the operative framework of classical self - non-self discrimination appears as a systemic emergence and cannot, without contradiction, be purely and continuously derived from descriptions of the elemental interactions. Of course, this compatibility between the idiotypic network hypothesis and the self - non-self discrimination does not mean that those two notions validate each other: that is another problem.

The distinction of levels is somewhat artificial (or, more precisely: conventional), but the same is true for the constitution of any scientific object in general: as they are abstract objects, the ESSENCE of them cannot be the entire set of their possible characters. In that sense (but not only in that sense), scientific object is contextual and consciousness is partially, but directly, constitutive of that object. For example, "horror autotoxicus" concerns an individual. The non-destruction of this individual by itself is involved in the abstract idea of "individual" when only this individual is considered. In that context, disease appears as an anomaly, as the death of the individual results from dysfunction. Physiology (what characterizes the individual as a self-maintained being) is opposed to pathology. If we now consider a society composed by "these" individuals, disease affecting some of them can be considered as constitutive of the physiology of this society, and individual death represents a physiological phenomenon at this social level. In other words, the individual is not exactly the same in both cases, even if it has the same name, and the observer is directly involved in this constitution. The object (here the individual) is contextual and conventional (we must explicitly or implicitly refer to a common usage). Of course, this brief remark cannot account for the complexity of object constitution and observer involvement. This focusing on increasingly larger objects could give the idea of a growing series of sets, each one containing the previous. Let us just say that this is not our conception and we must, for instance, take into account the "strange loops" of Hofstadter (2). Each set is constituted by its elements and is constitutive of them. The notion of "self" (at least, what we will call later self as an a priori category) being necessary to constitute an entity which at the same time gives birth to the applied notion of self, the heterogeneity which begins to appear must concern equally both self and the entities utilised.

If we refer to common usage, it seems that "self" is conceived as a set of de jure listable entities (only their great number prohibits establishment of this list in fact). Non-self, in contrast, is the negation of self (as already indicated by its designation). However, are this self and non-self in the same terms ? We will not normally say that a cloud or a lake is non-self for the mouse, but not because it is wrong. Focusing on an entity (i.e., the mouse) when speaking about self implies at once, TO SOME EXTENT, a sort of positive constitution of the non-self's domain. Nothing FORMALLY prevents us from considering a screw as mouse non-self, but, in immunology, we do not expect to get anti-screw antibodies even if administrated in complete Freund's adjuvant. Focusing on the "self" VERY ROUGHLY determines the domain of relevance of its negation. This implies that we cannot use here the very classical postulate: non non-self = self (at least without added constraints). Non-self being negatively constituted, it could be a set of de jure listable entities only if one could postulate that non non-self = self and if the self was represented by a set of DE FACTO listable entities (and not only de jure).

Before continuing, let us follow up this, by focusing on the mouse (as a general abstract mouse). Let us consider that the relevant constitutive entities here are biochemically defined molecules. A molecule (a class of molecules) which we can extract from an animal other than the mouse is "self - mouse" in these terms. In this case, the universe of the non mouse contains self - mouse molecules. We CANNOT therefore identify the universe of non self with the universe of the non mouse, when we MUST assimilate the universe of self with the universe of the mouse. Non self is effectively negatively determined.

In immunology, if this analysis is theoretically identical, determination of the self that could eventually allow us to establish de jure a list of self antigens requires, at least in principle, the immunological non responsiveness, which means that this self would be "defined" by ... the negation of the non self. Therefore, according to what we have previously said, we do not have, in this case, any positive reference.

This allows us, in this context, to ask more precisely the question which interests us here: the difference between both situations above (self of an animal in terms of biochemically defined molecules and the immunological self) essentially consists in that it is no longer the observer alone who erects a discriminative system according to his goals, but the observer who establishes a system which "by itself" is able to operate this distinction. By itself means that, being only in part constituted by the observer, this system does not have, a priori, any necessity to operate a distinction in extactly the terms of the observer. In other words: does the immune system discriminate the observer's objects as a function of the observer's categories ? If not, there would be heterogeneity of the objects as they are constituted and at least partial inadequacy of them in the study of immune system.

In order to delimit somewhat this question, we have to pay some attention to the notion of self in a general manner. We will then try to see how immunological objects are constituted, restricting ourselves to antibody - antigen couple to retrieve, at this very practical level, the heterogeneity present at a general one.

SELF AND NON SELF AS A PRIORI CATEGORIES

Until now, we have used self and non self conforming to general usage which permitted us to evoke the problem of the constitution of a scientific object. This problem represents for us one of the major questions in modern biology, and it is certainly out of question to tackle it in a consistent manner. Nevertheless, we have to briefly come back to it now.

It is sufficient to ask if an object, for instance a given molecule, is self for a given animal, to realise that we have already in mind two categories: self and non self, which allows us to ask the question.

Ambiguities or difficulties that can be encountered when trying to qualify one or another entity do not depend on the fuzzy nature of these categories, but arise from the uncertainty as to the relationships that a given object has with one of these categories.

These categories are also a priori, pure concepts in the sense that they are almost out of context (their context is heterogeneous with contexts of application) Without getting involved here in understanding or establishing the philosolophical nature or status of those categories, we can simply say that, at a certain moment, we ask the question of self - non self discrimination in a CERTAIN DOMAIN. To do so, we dispose of two categories which must be present before any discussion concerning that subject (which itself requires these categories) and which are necessary to qualify (in a heterogeneous manner) objects about which we want to speak. Being the very minimal condition of expression and knowledge, those categories cannot be knowledge by themselves.

The expression "a priori" is obviously very loaded, but the reference to Kant is, for us, limited. We do not want to state that these categories are not a historical emergence from our experience, constituted by it and constitutive of it. The main point is that, HERE AND NOW (hic et nunc), two categories appear as distinct and a priori, and are essentials and not the self or non-self of something. The applied notions of self and non self we have previously discussed require these a priori categories in order to be expressed and conceptualised, but we have seen that they differ from a priori categories if only because they are neither distinct nor exclusive. "Self" (and "non self") refers, through a form which is necessarily identical, to concepts which are profoundly heterogeneous in so far as their nature goes, but which all require for their constitution these a priori categories, empty of knowledge. Indeed, it appears impossible as we said at the beginning, to bring a procedure in operation allowing us to decide if one object is self or non self without using, inside this procedure, the very concept of self. Any object, any entity requires the use of this concept in its very constitution, whether it concerns an individual, a group, a class etc.

Indeed, if, for instance, we consider an immunoglobulin or an enzyme of an individual animal or of a "given" group of animals, these molecules appear as obviously "self" for this animal or this group, but this "obviously" refer to a pre-existing complex use (hic et nunc) itself variable. So the genetic reference (what is coded on genes tends to appear nowadays as the main reference in the field) reinforces and modifies the notion of self, but cannot create it. In other words, the categorical self is already necessary in the establishment of this genetic reference. It suffices then, to improve the proposition's coherence, to consider this genetic reference (a historical emergence) as a major reference in the constitution of living objects and relations used in this domain ; even, at the limit, to create an operative tautology by giving a normative definition of self based on the genetic characters of the considered living object. As convenient as this extreme solution may appear, it does not erase the historical conditions of emergence of this normative definition, and such a procedure, indeed perfectly justifiable in the restricted context of a precise act, canalises the evolutional capacity of the original notion and does not solve the problem of heterogeneity. This also has consequences that we cannot dwell on here.

In summary, it is self which, for us, is self (and this includes a broad zone of necessary uncertainty) and, unless we state that there is no other mode of existence for the physical world than its description (which we do not state), there is no necessity for coherence between distinctions made by us concerning diverse objects and distinctions performed by a system established by us. In these conditions, in cases where some homogeneity seems to appear (for instance, when, from molecules described in terms of biochemical structures, one can reach a certain degree of prediction as to the immune system's behavior "in response" to those molecules), these cases are as susceptible to questioning as those where these predictions fail. We are far from basing the immune system on its ability to distinguish self from non self.

ANTIGEN - ANTIBODY

The heterogeneity of objects as we discussed above must be found at every level of the description. If an antigen is classically presented as being a molecular class (characterized by its biochemical structure and generalized in those terms), description of antigen - antibody relationships imposes the creation of what seems to be often conceived as a sub-entity: the epitope.

This subject has been approached by ourselves (3) and evoked by Kristian Hannestad during this meeting. A similar subject has been previously discussed (4). Briefly: antigen and antibody, in so far as immunological qualitites based on molecules, are mutually constituted. We do not have any procedure for predicting any of these qualities. As K. Hannestad pointed out during the present meeting, a molecule does not have epitopes. This quality is brought to the molecule by a paratope, which determines the place, the size, the molecular components implicated in the linkage etc. A paratope, to some extent (but to some extent only) concretises a possibility that can be expressed by a given molecule (to some extent only, because this possibility can be eventually formulated only after at least one realization of that epitope...). Once this epitope is realised, it always remains possible for the same given molecular class and, in that sense, a certain linkage exists between these heterogeneous qualities. It is not necessary to insist on the fact that very different biochemical classes of molecules can react with the same mAb. If different mAb can have different affinities for the "same" epitope, it means precisely that the epitope is not the same.

Surprisingly, the same type of reasoning applied to the paratope seems to be less easily admitted. Indeed, one can accept the idea that immunoglobulin molecules, in general, possess at least one special structure realised at the level of their variable regions which favor their linkage with other molecules. Take the case in which we say that each antibody possesses a hollow site particularly suitable to entertain numerous linkages with part of another molecule. Even in this case, the paratope only constitutes a restriction in space for realization of the "possible". Nothing indicates that in the different linkages that this paratope can constitute with antigens, the same constituents of the paratope are involved. Now, only the set formed by these linkages is relevant to characterize an immunological paratope. The paratope as such happens to be determined by the epitope, with some external constraints, for instance its maximal size. There is indeed here co-determination between epitope and paratope, **which cannot be considered as entities.**

If we consider a multireactive mAb, such as those produced by certain lymphocytes, which bind KLH, several albumins, several cyto-skeletal proteins, TNP, hemoglobin and so on, the traditional reasoning would lead us to say that all those molecules have a common epitope or that this mAb has different paratopes... (can it be that this traditional conception of epitopes beinq considered as entities is involved in the immunology community's reticence in accepting such multireactive mAB ?).

These multireactive mAb's are also said to be (at least for many of them) "anti-self", because they bind to self molecules. Can we assume that the epitope-paratope set concerned in all the situations where one can show a binding on molecules considered as "self" represents a homogeneous field in which all propositions concerning this characteristic would possess a general value ? The heterogeneity of the objects corresponds, of course, to a heterogeneity of concepts, and what is communicated through formally identical words does not necessarily possess a common meaning in the different contexts of their use. The notion of self - non self discrimination is born in the immune response context and, until now, it keeps track of its origin. As we have underlined at the begining, not only the immune response in so far as it is a systemic emergence of a reactivity cannot be reduced to merely the behavior of individual components of the system, but also a priori, it cannot constitute the sole expression of the system and cannot even be considered as the paradigm of the expression of the immune system. Nothing which is explicitly or implicitly linked to the traditional notion of "self" has a necessary linkage with the same word "self" used in a different historical context. So, because of the heterogeneity that we have underlined and very partially described, we do not see why, without any other reason, the fact that a mAb produced by a "spontaneously" activated B cell is able to bind to a molecule considered as "self" could **necessarily** contradict the idea, born and expressed in another context, according to which anti-self antibodies producing cells see their expression suppressed in the normal animal (whether this idea is "true" or not does not matter here).

These molecular "selves" arising from different immunological contexts do not NECESSARILY have to be common. If experimentally we find a certain degree of community between these different "selves", then we must ask what makes this possible and must not consider it as a necessary starting point. It seems to us that, for instance, it would be fruitful to reconsider in these terms notions like internal image and allotype.

We must now try to avoid some possible misunderstandings. Our intention is not to reject any use of the notion of self nor to judge the use of a molecular nomenclature to refer to heterogeneous objects. We do not want to have a normative position. Our goal is to delimit some aspects of these very important notions in immunology, and, especially to try to open them up. Heterogeneity seems to us to be a key concept in science. It is not per se a reason to reject something. What is irrelevant is to work with heterogeneous objects believing that they are homogeneous.

This previous analysis is obviously too short and too limited to sufficiently precise the status of the observer, but it is maybe already sufficient to be careful about the eventual use of semiotic methods in an immunological context. This warning has been given already by Umberto Ecco at the beginning of this meeting. As signification appears in the context of human consciousness and is used in that very context, even if this signification covers an enormous amount of unknown, it immediately has some necessary coherence with the language used to study it. The use of signification has some immediate relevance. In the study of the immune system, we are obviously concerned with semiotic as immunologists, but it seems to us that, from our previous analysis, signification has not this necessary relevance to the immune system. If the source of meaning has a different status in semiotic and in immunology, we do not see how one can expect to use one in the field of the other in a coherent manner. The point is not that problems are easier in one field than in the other, but that problems are different.

REFERENCES

1. P. Ehrlich, On immunity with special reference to cell life. Proc. Roy. Soc. B. 66: 424-448 (1900)
2. D. Hofstadter, "Gödel, Escher, Bach: an Eternal Golden Braid". Basic Books, Inc., New York (1970)
3. F. Jacquemart, Réseau idiotypique. In "Les théories de la complexité", Actes du Colloque de Cérisy la Salle, Juin 1984, Le Seuil, ed., Paris. To be published.
4. N.K. Jerne, J. Roland and P.-A. Cazenave, Recurrent idiotypes and internal images. EMBO J. 1: 243-247 (1982).

GENERAL PRINCIPLES OF COMPLEX BIOLOGICAL SYSTEMS OPERATING IN IMMUNOLOGY. SELF-RESPONSES MIGHT DEFINE THE BOUNDARIES OF THE DEVELOPING IMMUNE SYSTEM

C. Martínez-A., A. de la Hera, M.A.R. Marcos, C. Márquez, M. Alvarez de Mon and M.L. Toribio

Centro de Biología Molecular, C.S.I.C., Universidad Autónoma, Cantoblanco, 18049 Madrid, Spain

Correspondence: Dr. C. Martínez-A. Centro de Biología Molecular Universidad Autónoma, Cantoblanco, 28049 Madrid, Spain

> Occidental man's repeated discovery of instances and inability to perceive the underlying principles demonstrates the rigidity of epistemology.

There are two characteristics of the Immune System, common to complex biological systems, that have to be mentioned together, that is autonomy and death. Autonomy, from the Greek autos (self) and nomos (law) meaning literally "control of the self", is provided by the recursive structure of the system. Death is a consequence of the entity's being made of multiple parts. In death, these parts are disassembled or randomized and consequently result in the destruction of autonomy. However, if mere survival is the objective, then the harder rocks have to be placed near the top of the list as those most successful among macroscopic entities. Therefore, the simple tautology of natural selection, stated as those descriptive propositions which are true for the longest time remain true longer than those that become untrue sooner, needs to be taken into larger perspective. It is, in fact, possible that rocks resist change unchanging, while living things escape change by correcting change or by incorporating continual change into their own being (1). Stability may be achieved either by rigidity, by continual repetition or by some cycle of smaller changes through continuous return to a status

NATO ASI Series, Vol. H23
The Semiotics of Cellular Communication in the Immune System
Edited by E.E. Sercarz et al.

quo ante (2).

In addition, the Immune System is capable of purpose and of choosing the way of its self-corrective possibilities. It is influenced by "maps", never by territory, and is therefore limited by the generalization that its receipt of information will never prove anything about the outself or about itself. Beyond that, the system will learn (by remembering) and remember, and will be capable of joining with other similar systems to make still larger wholes.

1.- PROPERTIES OF COMPLEX BIOLOGICAL SYSTEMS. RELATION AND APPLICATION TO THE IMMUNE SYSTEM.

Some of the categories that should be mentioned in association with general properties applied to the Immune System are described below. We hope they will help us to understand the "languages" which the components of the Immune System used in their interactions.

A.- Complex biological systems are aggregates of interacting components.

Although it seems obvious and unnecessary to mention this point, it is, however, worthwhile to say that even when some parts of an aggregate may themselves be regarded as systems, there is always a lower level of division such that resulting parts, when considered separately, lack the complexity necessary to achieve the criteria for being "complex". It is our interpretation that the "complex" results as a consequence of interactions of differentiated parts and, therefore, wholes as constituted by such combined interactions.

What are the relationships between parts? How do parts interact to create a biological complex? These questions are still to be debated. The interaction between parts or components is triggered by differences. In the material universe, we are commonly able to say that the cause of an event is some force or impact exerted upon some part of the material system by some other part. Besides, in the Immune System, it takes a relationship, either between the parts or between a part at time 1 and the same part at time 2, to activate some third component which we may call the receiver. What the receiver responds to is a difference. In that way, difference, defined as the nature of relationship, is not located in time or in space (3).

To produce differences, there must be at least two entities such that the difference between them can be imminent in their mutual relationship. Such new differences can now be represented as a difference inside some information-processing entity that constitutes the "complex". Furthermore, the difference between the information provided by the two entities and that provided by the other is itself information of a different logical type. There is an answerable question about the nature of those "necessary" two things that between them can generate the differences which become information by making a difference.

B.- <u>The essence of complex biological systems requires circular chains of determination</u>.

The idea that circular causation is of very great importance was first generalized by Norbert Weiner (4), applying mathematics to nonliving systems. In a circular system, it is very easy to visualize any biological process as being interconnected among its parts as well as connected to the outside, and consequently, any effects of events at any point in the circuit can be carried throughout to produce changes at that point of origin.

It is also important to point out that there might also be circuits of causation which would contain one or more negative links and might, therefore, be self-corrective. Such self-corrective circuits would provide the possibilities for modeling the adaptive action of the organism. In fact, many self-corrective systems are known: epidemics and wars can be visualized as such in population growth.

Complex systems, however, still require being characterized by time constants to be considered as a whole. After all, the entire process occurs in time. In this way, changes in the behaviour of the parts can be noted whose magnitude and timing were determined by forces and impacts between the separate components of the circuit. So a change in A determines a change in B and so on. When the description reaches the arbitrary place at which it began, there is a sudden change. The description must now compare change with change, and use the results of that comparison to account for the next step.

C.- Repetition of states defines the self-reference and the stability of these systems.

A system or structure which has the capacity of self-awareness will be called a "unitary structure" in the sense that knowledge about it does not begin in external sources. Such a unitary structure exists from itself, even if this "self"-existence also requires a continued addition of something "outside" the unitary structure (4). The unitary structure will then consist of a number of repetitions of some basic elements. In this structure, a phenomenon, or event, can only be known or identified as something singular, or specific, via its relationship with its own repetition. Hence a structure in a unity of self-relationship, or a unity of self-repetitiveness is the knowledge of a phenomenon with the aspects of the knowing mind which are repetitions of the phenomenon.

2.- BASIC MECHANISM OPERATING WITHIN THE IMMUNE SYSTEM

The previously defined criteria, together with the existence in the Immune System of: a) internal activity associated and required for self-responses; b) two basic components, B and T lymphocytes; and c) continuous repetitions of events that might appear early during the ontogeny of the System, constitute the basis for the existence of the mechanisms described below.

A.- Recursive behaviour.

There is substantial evidence today that the Immune System displays a certain degree of natural internal activity characterized by the existence of effector cells in the absence of conventional immunizations, and also present in animals kept in germ-free conditions and fed on chemically defined, low molecular weight diets (5,6). Furthermore, both B and T lymphocytes participate in such an activity, which most likely constitutes one of the essential characteristics of a system in continuous dynamic steady state equilibrium (7,8). Consequences of this activity are, for instance, the normal values of immunoglobulins in serum containing a physiological concentration of antibodies directed against self-components (9). These autoantibodies, together with the self-recognizing T cells, constitute the basis for self-recognition as a principle responsible for the initiation and regulation of the immune responses (10). These principles, initially thought to function through recognition of major histocompatibility (MHC) antigens and the idiotypes on immunoglobulins, have been extended to all the polymorphic structures of the Immune System (immunoglobulins, Ig; T-cell receptor, T_{CR}; and MHC) as well as to all other components of the internal environment of self in an immunosomatic perspective (11).

One of the basic principles operating in the initiation and maintenance in these activities appears to be the recursive mechanism between B and T cells operating in the selection of immune repertoires. Thus, T-cell repertoires originally selected on the basis of the confrontation of T-cell precursors with self-MHC are afterward selected positively or negatively in the periphery as a consequence of the interaction with another major selective force, that is the idiotypes expressed either on the immunoglobulins or on the T_{CR}. These interactions, which are now on the way to being understood, are also used at the same time to explain one of the most paradigmatic problems in recent immunology, that is idiotype sharing between B and T lymphocytes (12). Today, this whole finding is explained as a consequence of regulatory interactions occurring most likely in the periphery, mediated by idiotypic complementarities between immunoglobulins and T cells (8). Besides, evidence of the emergence of a clonotypic network as soon as early thymocytes acquire T_{CR} (13,14), together with the mimicry of class II MHC antigens expressed on T_{CR} (15), will set up the basis for the mechanism operating in the selection and maintenance of T-cell repertoire on the basis of the repertoire expressed by the immunoglobulin and T cells (16).

On the other hand, there is compelling evidence that actual repertoires of antibodies are in part selected on the basis of helper cells. In fact, data has been produced indicating that helper T cells are responsible for the clonal expansion and immunoglobulin secretion of B cells responsible for the natural internal activity. Furthermore, the MHC control of certain immunoglobulin idiotypes, together with results showing that idiotypic systems defined on T_{CR} and also expressed on immunoglobulins are strictly under T-cell and MHC control in order to be expressed in the latter, clearly speaks in favour of the recursive T-B interaction as responsible for the nature of the selected B cell repertoire (17).

We should expect then that a given pattern of recognition, selected on the basis of MHC and immunoglobulins, will in turn select a given B-cell repertoire which will reflect the influence of the three gene families, mainly, Ig, MHC and T_{CR}, implying that this interaction is directly responsible for the recursively selected repertoires of both B and T-cell compartment.

B.- Hypercycling mechanism.

In vivo injection of nanogram amounts of a certain idiotype induced increased circulating levels of the same idiotype (18, 19). This property of "autoreproduction" is consistently found in natural antibodies present in the serum. Furthermore, this property is under immunoglobulin allotype control, under the influence of MHC and mediated by T cells since this principle does not operate in nude mice (19).

More recent experiments have defined two major populations in the Immune System, characterized as being small or activated (5,6). Each subset engages in immune responses, the former if stimulated by external antigens, and the letter by displaying effector activity. A characteristic of members of the activated set populations is that they display high connectivity among the other components of the same set of cells, property that is not present in the pool of small resting cells.

Considerable evidence has been accumulated suggesting that an idiotypic network organization which is functionally relevant is primarily or exclusively present in the compartment of activated lymphocytes (20). Besides, only members of this set display this characteristic of the hypercycling mechanism, which most likely results as a consequence of the repertoire selection imposed upon the network interactions. The existence of this hypercycling has been extended into B-cell populations displaying the $Ly\text{-}1^{+}$ phenotype. In fact, permissive mouse strains, like CBA/N, allow "autoreproduction" of $Ly\text{-}1^{+}$ B cells normally lacking in these mice, after BM transfer from syngeneic CBA/N in which has been induced a chronic graft versus host disease, characterized, among other things, by having Ly-1^+ B cells, are only representative of the B cell compartment (21,22).

Recently, Ly-1^+ B cells have been distinguished by their unique role in immune physiology and diseases. These cells, initially characterized by their novel surface expression of the Ly-1 antigen, are most prominent early in ontogeny and they have a restricted time distribution and precursor origin (23). They are the producers of autoantibodies in normal and autoimmune mouse strains, provide a novel and stimulatory influence on antibody production by other B cells and impose selective expansion of T cells, due in part to the production of anti-idiotype antibodies. T cells with characteristics similar to Ly-1^+ B cells, that is displaying preferential self-reactivity, are selectively expanded in young mice and in relatively high percentages in the peritoneum; they are also in the process of being characterized (24 and unpublished results). Ly-1^+ B-cell expansion is most likely due in recipient CBA/N mice to selective stimulation of complementary T cells in the host, that in turn will be induced by the complementary Ly-1^+ B cells.

CONCLUDING REMARKS

This mechanism enables the Immune System to be organized in a closed, circular organization based on an idiotypic network (25,26) in which self-reactivity to all the somatic components is a characteristic of the naturally or internally activated set of T and B lymphocytes. Functionally autonomous and independent of external challenges, the circularity of its organization provides stability and recursivity in repertoire selection where, for each component, complementarities and mimicries are both to be found.

On this basis, we can conclude that it might very well be that self-antigens exert pressure on the shaping of the immunoglobulins and T-cell receptor V genes, anti-idiotypic responses and mimicry against anti-self. The ontogenic development of the Immune System is based therefore on the initial self-recognition and consequent anti-self external expansion, the complementary interactions between B and T cells being, subsequently, the key elements in the constitution of the system. However, the principle operating in these interactions is based on continuous repetitions, and therefore, nothing is learned but remembered. Idiotype sharing between both sets of lymphocytes can constitute the reference for self-correction as well as for self-determination in the genetically determined incipient network.

ACKNOWLEDGMENTS

We thank M. Messman for her skillful care in typing the manuscript. The work described is supported by grants from FISS, CAICYT and the VW Foundation.

REFERENCES

1. J.R. Young. 1961. Doubt and certainty in science. Oxford University Press.
2. G. Bateson. 1979. Mind and nature. Fontana.
3. D.R. Hofstadter. 1985. Metamagical themes. Basic books.
4. N. Barraclough. 1980. Preology. Pergamen Press.
5. P. Pereira, E.L. Larsson, L. Forni, A. Bandeira and A. Coutinho. 1985. Proc. Natl. Acad. Sci. USA 82, 7691.
6. P. Pereira, L. Forni, E.L. Larsson, M.Cooper, C. Heusser and A. Coutinho. 1986. Eur. J. Immunol 16, 685.
7. R.R. Bernabe, A. Coutinho, C. Martínez-A. and P.A. Cazenare. 1981. J. Exp. Med. 154, 552.
8. C. Martínez-A., R.R. Bernabe, A. de la Hera, P. Pereira, P.A. Cazenave and A. Coutinho. 1984. Proc. Natl. Acad. Sci. USA 81, 4520.
9. S. Avrameas, B. Guilbert and D.G. Dighiero. 1981. Ann. Immunol. 132, 231.
10. C. Martínez-A., A. Coutinho, M.L. Toribio, P. Pereira, A. de la Hera, C. Márquez, M.A.R. Marcos. 1987. International Conference on Idiotypes. Academic Press, in press.
11. N. Vaz, C- Martínez-A. and A. Coutinho. 1984. Idiotype in Biology and Medicine. Academic Press, pp. 43.
12. K. Eichmand and K. Rajewsky. 1977. Contemp. Topic in Immunobiol. 7, 69.
13. M.L. Toribio, C. Martínez-A., M.A.R. Marcos, E. Cabrero, C. Márquez and A. de la Hera. 1986. Proc. Natl. Acad. Sci. 83, 6985.
14. A. de la Hera, M.L. Toribio, M.A.R. Marcos, C. Márquez and C. Martínez-A. 1986. Scand. J. Immunol., in press.
15. A.A. Augustin and G.K. Jim. 1983. Ann. N.Y. Acad. Sci. 418, 272.
16. A. Coutinho. 1980. Ann. Immunol. 131, 235.

17. A. Coutinho. 1982. Curr. Topics Microbiol. Immunol. 98, 113.
18. L. Forni, A. Coutinho, G. Köhler and N.K. Jerne. 1980. Proc. Natl. Acad. Sci. USA 77, 1125.
19. D. Holmberg, F. Jvars and A. Coutinho. 1983. Eur. J. Immunol. 13, 82.
20. D. Holmberg, A. Freitas, D. Portnui, F. Jacquemart, S. Avrameas and A. Coutinho. 1986. Immunol. Rev. 93, 147.
21. A. de la Hera, M.A.R. Marcos, C. Márquez, M.L. Toribio and C. Martínez-A. 1986. Submitted.
22. M.A.R. Marcos, A. de la Hera, M. Gaspar, C. Márquez, F. Mampaso, C. Bellas, M.L. Toribio and C. Martínez-A. 1986. Immunol. Rev. 94, 51.
23. G. Moller (Ed.). 1986. Immunol. Rev. 93.
24. H. Ishikawa and K. Saito. 1980. J. Exp. Med. 151, 967.
25. N.K. Jerne. 1974. Ann. Immunol. 125, 373.
26. N.K. Jerne. 1984. Immunol. Rev. 79, 5.

Towards Cognitive Immunology : G.O.D. is Infinite.

Nelson M. Vaz
Departamento de Bioquímica e Imunologia
ICB-UFMG 31270 Belo Horizonte MG Brasil

Immunosemiotics and cognitive immunology.

The meeting in Immunosemiotics was a rare opportunity to take a fresh look into immunological problems. It was possible, for instance, to define them as cognitive problems -i.e.: problems involving the interpretation, computation and generation of knowledge. It is true that only in a few moments during the meeting this possibility was clearly brought forth; but a few important statements were made in this direction, and I will use this contribution to reinforce this view. This will be an effort of the sort Francisco Varela has called "...the difficult process of addressing foundational issues at an unreceptive time."(1). And this will be so because we live in a tradition which, although usually invisible, influences our ways of delineating problems and our criteria of validation of hypotheses and results.

Networks, internal and external images.

In 1974, Jerne propose ideas "Towards a network theory of the immune system"(2,3). These ideas might have been interpreted as a first step in the direction of a cognitive immunology (4). But they were not. The emphasis was on the interconnectivity of immune events, and most of the effort was placed on understanding how cellular circuitry could help to regulate specific immune responsiveness. This drove "network immunology" away from its truly important potentialities. And this loss of momentum occured because, in their essence, network concepts are not compatible - i.e.: commensurate with the notion of specific immune responsiveness.

NATO ASI Series, Vol. H23
The Semiotics of Cellular Communication in the Immune System
Edited by E.E. Sercarz et al.

Immunoglobulins and T cell receptors (V-region bearing molecules) function not only as receptors(paratopes) for foreign elements (epitopes) but also bear determinants (idiotopes) which are bound by other paratopes of the same immune system. The same or similar idiotopes may be associated with different paratopes creating a complex web of connections throughout the immune system. The repertoire of idiotopes is imagined to be as large as that of the paratopes; thus, any foreign epitope may meet not only complementary paratopes, but also idiotopes which are similar to its own structure - i.e.:"internal images".

It would have been better to call epitopes "external images" of idiotopes. This would be an important change in perspectives, and not simply a play of words. The notion of internal images is representationist (1) in the sense that the external world of epitopes is "represented" by an internal world of idiotopes. The duality of worlds is an essential quality of the objectivist tradition in which we are immersed. On the other hand, the notion of "external images" has, or may be given a tinge of hermeneutic (interpretative) qualities, and has the advantage of "centering in the organism" the origin of immunological processes (4).

Instead of facing the sudden intromission of an epitope as a starting event, demanding a response, we might decide that what matters is its assimilation - as Jean Piaget would put it(5,6) - into ongoing processes. We could shift our attention from the interaction between foreign epitopes and specific lymphocytes, to the interactions of lymphocytes among themselves; from the local rules governing the operation of components of the immune system, to the search and understanding of the global rules of interaction among these components.

But this did not happen, and the attention remained focused on cellular circuits and specific immune responsiveness.

This is somehow ironic because Jerne himself has been telling us since 1955 (7) that immunological interactions are events of low-fidelity: each paratope interacts, or has the potential to interact with thousands of different epitopes (8) or idiotopes(2, 3). The interactions are reversible, changeable, replaceable by others. The multitude of different components and subcomponents utilized by the immune system in every one of its actions makes it irrelevant whether any particular component or subcomponent participates in them.

These ideas are favorable to the notion of an everchanging web of interactions among lymphocytes, but not to the notion of specific immune responsiveness. Actually, I want to propose that the traditional ideas of specific immunological responsiveness, and "network ideas" imply polarly opposite concepts. This is an insight that may be derived from the most characteristic aspect of immune phenomena - i.e.: the ceaseless generation of diversity (G.O.D.) represented in V-regions of paratopes.

G.O.D. is infinite.

The strategy of the immune system to deal with continuos contact with novel molecules - either absorbed from the environment or generated internally, for example, in decaying cells - is to "dilute" this newness in a much larger pool of internally-generated components, which is constantly renewed. I am proposing that through special mechanisms of gene modification, the immune system continuously produces a collection of subcomponents which, for all practical and computational purposes, and for the operation of the system itself, may be considered infinite.Externally -derived change cannot add to this already massive generation of diversity (G.O.D.).

In this organism-centered perspective (4) the goal is to elucidate the autonomous organization of the system (9-11) the pathways through which the endless emergence of internal newness is handled by the system, because external materials would not be perceived by the system as an special class of materials,but rather as aspects of ongoing internal changes. For the system, there would not be an inside and an outside - only its patterns of operation (4,9-11).

In this perspective, although the immune system does not make a distinction between internal and external materials, we, as observers of immune systems, are certainly free to make it. The way we chose to look into immunologic phenomena is highly specific; I am suggesting that the cross-connected, redundant manner by which the phenomena themselves occur in nature,is not specific.

Associative survival.

Studies on the life history of lymphocytes (12,13) show that a large proportion - actually, the majority - of lymphocytes produced in the immune system are discarded soon after their emergence. The small proportion of surviving cells apparently survive because they associate with other cells admitted previously into the system, and encounter essential growth and differentiation factors. The local rules of incorporation of new lymphocytes may resemble the rules proposed by Donald Hebb in 1949 , saying that the strength of a connection in a (neural) network increases on the basis of the simultaneous firing on both sides of the connection. Granted that the connected elements are globally interlinked in a network, simple rules such as Hebb's rule may generate complex patterns of global behaviour (4,11,14).

An important aspect of this organization is that, through the associative incorporation of new elements, the composition of the system embodies, in itself, previous actions of the system. Thus, the composition of the system cannot be dissociated from its history. Which types of emergent new components will be incorporated into the system is a function of its epigenesis (9-11). An important development in this direction has been the recent demonstration that Ir-gene low-responder mice can be turned into high-responders by immunoglobulin manipulation of the developing immune system(15).

Redundant versus specific.

If we admit that the immune system dilutes external newness into a much larger (infinite) pool of internal newness, and deals essentially with patterns of change, we can no longer hope to explain the fundamental properties of the system in terms of specific components and subcomponents, or their "repertoires". Although any of the interactions taking place in the system at each moment may be said to be "specific" in the traditional sense, it is no less true that any of them could be replaced by similar ones, which are actually taking place, or might be taking place in the same immune system. The fundamental aspect of these interactions, therefore, is not that they are specific, but rather that they are multiple, variegated, vicariant, redundant. We are biased by the stringent specificity of our immunological observation,and thus led to believe that the immune system itself operates through equally specific events.

If the system operates through patterns of recognition, we may have been directing our efforts in the wrong direction for quite a long time. An "artificial nose" operating on the basis of pattern recognition may recognize a wide array of chemical "odors" using only three types of bimetallic smoke-detectors as "sensors"(16). The variety of recognitions performed by the immune system may be much larger than the number of specific lymphocyte clones, because the same clones participate - in a degenerate, redundant manner - in many different patterns.

Since the immune system is endowed with mechanisms for the generation of diversity we all have been led to believe that this variety is necessary for the performance of "specific" items of recognition, when actually this diversity is necessary to ensure the vicariant, plastic quality of the patterns of change. Therefore, if "specific" is meant to signify a distinct, individualized thing, as the term idiotypic - from idios: person al, separate - implies, then "specific" is exactly what natural immunological phenomena are not. The immune system may be described as an artifact which generates a perpetual process of change, always recruiting a variety of different components in the performance of all its actions and recognitions.

Conclusion.

In adopting a radical reading of network ideas it is possible to see them as polarly opposite, instead of complementary to the notion of specific immune recognition. The recursive and redundant emergence of an immense number of components,which may coactivate each other in a cross-connected manner, may be seen as the fundamental characteristic of immune systems - rather than a simple device for the regulation of specific immune responses.

The network approach (2,3) which may lead into the study of the autonomy of immune systems(4,9-11) is not a theory, as clonal selection claims to be, but a new paradigm in which new problems are delineated and some of the old problems dissolve as meaningless; it is a wider bridge of contact with other areas of scientific endeavor in which immunology may find useful conceptual tools.

Bibliography.

1. Varela,F.J.(1986)Experimental epistemology:Background and future.In Cognition et Complexité,Cahiers du CREA 9:107-122
2. Jerne,N.K.(1974)Towards a network theory of the immune system Ann.Immunol(Paris)125C:373-395.
3. Jerne,N.K.(1976)The immune system: a web of V-domains. Harvey Lectures 70:93-110.
4. Vaz,N.M. and Varela,F.J.(1978)Self and nonsense:an organism-centered approach to immunology. Med.Hypotheses 4:231-257.
5. Piaget,J.(1967)Biologie et Connaissance, Gallimard, Paris.
6. Piaget,J.(1976)Comportement:Moteur de l'Evolution,Gallimard, Paris.
7. Jerne,N.K.(1955)The natural selection theory of antibody formation. Proc.Natl.Acad.Sci.U.S.A.41:849-860.
8. Inman,J.K.(1978)The antibody-combining region:speculations on the hypothesis of general multispecificity.In G.A.Bell,A.S. Perelson,and G.H.Pimbley,eds.Theoretical Immunology,Marcel Dekker,New York, p243-278.
9. Maturana,H.R. and Varela,F.J.(1980)Autopoiesis and Cognition: The realization of the Living. Reidel, Amsterdan.
10. Varela,F.J.(1979)Principles of Biological Autonomy.Elsevier, North-Holland,Amsterdan.
11. Maturana,H.R.and Varela,F.J.(1987)The tree of knowledge: the biological basis of human uderstanding.New Science Library , Boston.
12. Freitas,A.A.,Rocha,B.and Coutinho,A.(1986)Lymphocyte population kinetics in the mouse.Immunol.Rev.91:5-37.
13. Freitas,A.A.and Coutinho,A.(1981)Very rapid decay of mature B lymphocytes in the spleen. J.Exp.Med.154:994-999.
14. Varela,F.J.(1987) this volume.
15. Martinez,C.-A.,Marcos,M.A.R.,Pereira,P.,Marquez,C.,Toribio,M. Cazenave,P.-A.and Coutinho,A.(1987) Turning (Ir-gene) low-responders in high-responders by antibody manipulation of the developing immune system. J.Exp.Med. (in press)
16. Persaud,K. and Dodd,G.(1982) Analysis of discrimination mechanisms in the mammalian olfactory systems using a model nose. Nature 299:352-355.

Supported by grants from CNPq/Finep/Fapemig(Brasil)

PARASITES AND THE IMMUNE RESPONSE: CROSSING THE SIGNALS

Rona J. Mogil
Department of Immunology
University of Alberta
Edmonton, Alberta T6G 2H7 Canada

INTRODUCTION

The co-evolution of parasites and their hosts has resulted in stable, species-specific associations. The most successful parasites are those which do not kill their hosts, but also resist their host's efforts to dispense of them. While hosts have developed novel immune mechanisms to control or eliminate their parasites, parasites have also adapted unique strategies of evading immune recognition while opting for a viable, albeit complex co-existence. The net effect ensures completion of the parasite's reproductive cycle and transmission. This state depends on the balance of the signals elicited by both the replicating parasite and the host's response to infection. The optimal state is achieved through mutualism; at the other extreme is death of the host with its associated parasites. However, an intermediate state more often succeeds in which a state of premunition (1) or concomitant immunity (2) occurs. Here we observe small numbers of parasites surviving despite aggressive host immunity, thereby ensuring both host survival and eventual parasite transmission. The means by which parasites evade host defense mechanisms can be construed as "lying" to the immune system, that is, creating a false sense of balance whereby host protective mechanisms may even serve to stimulate parasite growth. How then do infectious agents subvert host immune responses to their advantage?

The most common mechanisms described involve the induction, acquisition or mimicry of host transplantation antigens. Immature erythrocytes infected with murine plasmodia have been demonstrated to express elevated levels of Class I MHC antigens on their surface (3). Similarly, cells infected with Adenovirus 2 express elevated Class I (4). In contrast, schistosomula

NATO ASI Series, Vol. H23
The Semiotics of Cellular Communication in the Immune System
Edited by E.E. Sercarz et al.

adsorb soluble host Class I transplantation antigens onto their surface tegument soon after penetration of host tissues (5). Bacteria belonging to the Klebsiella species possess unique antigens which bear serological identity with Class I HLA-B27 epitopes (6). It has been proposed that MHC polymorphism evolved as a consequence of selection pressure by infectious agents (7), thus forming a basis for Ir gene-mediated enhancement of the ability of T lymphocytes to recognize and respond to an inestimable variety of antigens. Since MHC Class I transplantation antigens may be recognized as self, nonself or modified self, they comprise a reference framework for cell-mediated defense mechanisms. Such self signals may play an important role in the immune response to erythrocytic infections caused by the malarial parasite Plasmodium.

THE INTRACELLULAR HABITAT INFLUENCES PARASITE IMMUNOGENICITY

Erythrocytic infection with plasmodia can be conveniently studied in the mouse. Plasmodium yoelii 17X (PY 17X) occurs as two variants which differ by a single gene trait (8). The lethal (PY 17X-L) variant causes fulminant infections characterized by rapid exponential growth with parasitemia levels of greater than 80% of total circulating red blood cells prior to death of the host. Animals succumb 7 to 10 days after initial infection. In contrast, the nonlethal variant (PY 17X-NL) replicates more slowly with parasitemia levels peaking at 10 to 30% after which parasites are cleared from the peripheral blood. Within 3 to 4 weeks after initial infection animals recover and are refractory to reinfection with both the nonlethal and lethal variants of PY 17X. Curiously, the maturity of the predominant red blood cell infected by these two parasites differ. PY 17X-L infects mature erythrocytes whereas PY 17X-NL parasites are found within immature red blood cells (reticulocytes). Furthermore, early after infection with nonlethal PY 17X-NL, reticulocytosis is stimulated, thereby increasing the availability of uninfected target cells. This does not occur during lethal infections suggesting that the hematopoietic response is intimately tied with recovery. Indeed, if one induces a hemolytic anemia with phenylhydrazine during lethal PY 17X-L infection a reticulocytosis occurs, parasites invade immature red blood cells and infections are controlled (3). These recovered animals are immune to rechallenge with lethal and nonlethal PY 17X. This illustrates that the intracellular habitat influences immunity to blood-stage malaria. However, immunity is T cell dependent. T cell-deprived animals succumb to infection despite the presence of parasitized reticulocytes (PRETIC) (9).

Reticulocytes differ from mature erythrocytes by their higher level expression of Class I MHC antigens. Class II MHC antigens are not detectable on any stage of circulating red blood cell (10,11). When murine reticulocytes are infected with PY 17X-NL, levels of Class I K and D antigens are elevated (3). The extent of enhancement varies with the responder phenotype of the host to the parasite. Parasitized reticulocytes (PRETIC) from high responders possess the highest levels whereas PRETIC from low responders bear low levels (3,12). Thus both host responsiveness and parasite immunogenicity are intimately tied with Class I antigen expression. By analogy, lethal PY 17X parasites within reticulocytes become more immunogenic perhaps due to association with Class I antigens. This may induce a conformational change in the relevant antigen(s) which are shared with PY 17X-NL.

ROLE OF CLASS I MHC IN PROTECTION

Protective parasite-directed antigens can be coprecipitated from infected reticulocytes with either anti-Class I or anti-β_2 microglobulin antibodies indicating that these antigens are associated with Class I glycoproteins on the cell surface (12). *In vivo* protection with such host-parasite antigen complexes suggest that the presentation of parasite antigens in the context of Class I MHC constitutes an Ir gene phenomenon. However, immunization of allogeneic recipients with PRETIC from high responder donors has revealed that the induction of immunity is not H-2-restricted; that is, parasitized red blood cells (PRBC) of any genotype can immunize. Similarly, sensitized T cells proliferate in vitro when stimulated with PY 17X-NL PRBC from either syngeneic or allogeneic donors. However, the induction of immunity by PY 17X-NL PRBC vaccines is enhanced by the presence of elevated levels of Class I MHC antigens. Immunization with allogeneic PRETIC bearing high, intermediate and low levels of Class I antigens of the same haplotype clearly shows that the most immunogenic PRBC also bear more Class I. Furthermore, low responder phenotype correlates with the relative inability to recognize and respond to high levels of Class I (12). Thus, host immune responsiveness reflects both a permissiveness to allow parasite influences on MHC antigen levels as well as the ability to perceive such antigens. Athough Class I expression is important, immunity induced by some plasmodial antigens expressed on infected red blood cell membranes probably occurs independent of H-2 restricted interaction of PRETIC Class I MHC with the T cell receptor. This contrasts the situation of the recognition of bacterial and viral cell-associated antigens in which T cell recognition does occur in the context of Class I MHC genes (13,14). Therefore, Class I MHC may send different signals between parasites and the immune system. The role of elevated Class I MHC in the induction of anti-plasmodial responses continues to be investigated.

PARASITE EVASION: CROSSING THE SIGNALS

Why should a successful parasite exploit, or stimulate, host expression of Class I genes if this event subsequently activates protective anti-parasitic immune responses? Emerging evidence suggests that Class I gene products are often associated with nonimmunological functions. For example, a variety of hormone receptors which bind insulin (15), human growth factor (16), epidermal growth factor (17) and glucagon (18) have been observed to noncovalently associate freely with Class I molecules on target cell membranes. As suggested by Ohno (19), Class I molecule-β_2 microglobulin complexes may function as cellular binding sites for organogenesis-directing proteins. Such associations may represent a more primitive role for MHC, representing a basis for the manner in which antigen and MHC interact with the immune system.

In this regard, an antigen which associates intracellularly with Class I might enjoy coordinately increased expression when Class I gene products are induced at a higher rate. From these observations it becomes clear that it may be fortuitous for intracellular parasites to insert their own antigens into cell membranes in association with Class I. These may in fact be receptors for substances required by the parasite for intracellular growth and development. For example, plasmodia lack their own carbohydrate reserves and must obtain glucose from host plasma to meet their energy requirements (20). Thus, elevation of insulin receptors may aid in elevating intracellular glucose pools. Alternatively, such parasite-Class I complexes may activate production of lymphokines and cytokines which may stimulate the growth of the target cell population. Indirect evidence in support of this was obtained by immunizing mice with purified PY 17X-NL-Class I antigen complexes. Within three days after immunization a reticulocytosis occurred. Therefore, elevation of Class I expression with concomitant elevation of parasite-directed antigens affects the hematopoietic response to the malarial parasite's (as well as the host's) advantage. During an ongoing infection secretion or shedding of parasite-Class I complexes from PRETIC membranes into the peripheral blood circulation would also accomplish the same effects.

THE MALARIA HYPOTHESIS REVISITED

Resistance to malaria depends on the ability of the host to strategically adapt to maintain homeostasis. In this context, the exploitation of the immune system by plasmodia may be balanced by host genetic traits involving red blood cell disorders and hemoglobinopathies. The prevalence of lethal genes such as sickle cell, thalassemia and G6PD deficiency in African, Mediterranean and Asian populations has been attributed to the continual selective pressure of P. falciparum malaria (21,22). Epidemiological studies have revealed that in normal African children, immunity to lethal P. falciparum is not acquired until late in childhood. In contrast, children with G6PD deficiency or sickle cell trait develop immunity much earlier in life (22). A common hematological feature of these disorders is the presence of a reticulocytosis due to erythrocyte destruction or oxidant stress (R. Mogil, manuscript in preparation). One host-protective effect afforded by these genetic traits may be due to the availability of Class I expressing reticulocytes as the time of inoculation of sporozoites and after the hemolytic and hypoxic stress of erythrocytic infection. This would then confer an immunological advantage resulting in low grade chronic infections which are controlled. This balanced state of premunition, in which small numbers of parasites persist and continually stimulate host immunity, might be attained more easily and quickly in persons with red blood cell disorders than normal individuals due to their reticulocytosis.

CONCLUSIONS

Host immunocompetence involves a dynamic balance between parasite reproduction and host resistance. Unsuccessful parasites perturb this state and die with their hosts. However, co-evolution of infectious agents with their hosts has allowed mechanisms by which parasites may utilize host immune response gene products, such as Class I MHC, subverting normal cellular growth processes to their own advantage. These aspects of host-parasite interaction may result in greater benefit to the parasite. For example, regulation of Class I MHC expression on infected cells by adenovirus proteins may benefit intracellular viral growth perhaps by a mechanism similar to that described for plasmodia. Alternatively, the passive acquisition of host ABO blood group antigens and histocompatibility antigens on schistosomula surfaces may serve to mask target parasite antigens from immune attack. This masking might allow existing parasites to survive despite host immunity to reinfection.

These "lies" parasites tell effectively cross the signals of the communicating immune system. While parasites evolve more rapidly than their hosts, it is also apparent that human (immune response) genes have also undergone significant adaptive survival. For example, in response to falciparum malaria selection, human populations have attained balanced polymorphisms of red blood cell disorders. These heritable traits may offset parasite growth by facilitating the development of immunity at a younger age. The net effect is a balanced co-existence ensuring both host and parasite survival.

The immune system exists, in large part, for the control of parasitic infections. This obvious fact is often glossed over in our efforts to understand the signs and signals of the immune system. While much has been learned by the study of simplified systems (employing simple, nonreplicating antigens), the immune system has evolved under the influence of constantly evolving parasites. We may find that the immune system is even more complex than we expect and that its language reaches well into that of other physiological systems (such as that of the hemoatopoietic system discussed herein). Some parasites have learned to speak this language. A full understanding of host-parasite relations will come when we have learned to do the same.

REFERENCES

1. Sergent, E. (1963) Latent infection and premunition. Some definitions of microbiology and immunology. In Immunity to Protozoa. (Garnham, P.C.C., Pierce, A.F. and Roitt, I., eds.) Blackwell Scientific Publications, Oxford, p. 39.
2. Gershon, R.K., Carter, R.L. and K. Kondo (1967) On concomitant immunity in tumour-bearing hamsters. Nature 213:674-676.
3. Jayawardena, A.N., Mogil, R., Murphy, D.B., Burger, D. and R.K. Gershon (1983) Enhanced expression of H-2K and H-2D antigens on reticulocytes infected with Plasmodium yoelii. Nature 302:623-626.
4. Paabo, S., Kampe, O., Severinsson, L., Andersson, M., Fernandez, C. and P.A. Peterson (1985) The association between Class I transplantation antigens and an adenovirus membrane protein. Prog. Allergy 36:114-134.
5. Simpson, A.J., Singer, D., McCutchan, T.F., Sacks, D.L. and A. Scher (1983) Evidence that schistosome MHC antigens are

not synthesized by the parasite but are acquired from the host as intact glycoproteins. J. Immunol. 131:962-965.
6. Ogasawara, M., Kono, D. and D.T. Yu (1986) Mimicry of human histocompatibility HLA-B27 antigens by Klebsiella pneumoniae. Infect. Immun. 51:901-908.
7. de Vries, R.R.P., Meera Khan, P., Bernini, L.F., Van Loghem, E. and J.J. van Rood (1979) Genetic control of survival in epidemics. J. Immunogenet. 6:271-287.
8. Yoelii, M., Hargreaves, B., Carter, R and D. Walliker (1975) Sudden increase in virulence in a strain of Plasmodium berghei yoelii. Ann. Trop. Med. Parasitol. 69:173-178.
9. Mogil, R.J., Patton, C.L. and D.R. Green (1987) Cellular subsets involved in cell-mediated immunity to murine Plasmodium yoelii 17X malaria. J. Immunol. 138:1933-1939.
10. Brown, G., Biberfeld, P., Christensson, B. and D. Mason (1979) The distribution of HLA on human lymphoid, bone marrow and peripheral blood cells. Eur. J. Immunol. 9:272-275.
11. Spack, E and M. Edidin (1986) The Class I MHC antigens of erythrocytes: a serologic and biochemical study. J. Immunol. 136:2943-2952.
12. Mogil, R.J., Patton, C.L. and D.R. Green (1986) An immunogenic antigen of the nonlethal murine malarial parasite Plasmodium yoelii associates with Class I MHC gene products. Submitted for publication.
13. Kaufmann, S.H., Hug E. and G. de Libero (1986) Listeria monocytogenes-reactive T lymphocyte clones with cytolytic activity against infected target cells. J. Exp. Med. 164:363-368.
14. Zinkernagel, R.M. and P.C. Doherty (1974) Restriction of in vitro T cell-mediated cytotoxicity in lymphocytic choriomeningitis within a syngeneic and semiallogeneic system. Nature 248:701-702.
15. Samson, M., Cousin, J. and M. Fehlmann (1986) Cross-linking of receptors to MHC antigens in human B lymphocytes: evidence for selective molecular interactions. J. Immunol. 137:2293-2298.
16. Edidin, M. (1983) MHC antigens and non-immune functions. Immunol. Today 4:269-270.
17. Schreiber, A., Schlessinger, J. and M. Edidin (1984) Interaction between MHC antigens and epidermal growth factor receptors on human cells. J. Cell. Biol. 98:725.
18. Lafuse, W. and M. Edidin (1980) Influence of the mouse major histocompatibility complex, H-2, on liver adenylate cyclase activity and on glucagon binding to liver cell membranes. Biochemistry 19:49-54.
19. Ohno, S. (1977) The original functions of MHC antigen as the general plasma membrane anchorage site for organogenesis-directing proteins. Immunol. Rev. 33:59-69.
20. Sherman, I.W. (1985) Membrane structure and function of malarial parasites and the infected erythrocyte. Parasitology 91:609-645.
21. Haldane, J.B.S. (1949) Disease and evolution. La Recerca Scientif. (suppl.) 19:68-76.
22. Allison, A.C. (1984) Cellular immunity to malaria and Babesia parasites: a personal viewpoint. Contemp. Top. Immunobiol. 12:463-490.

'I SEE...

GRAMMARS, MACHINES, AND IMMUNOREGULATORY CELL INTERACTIONS. AN EXERCISE IN IMMUNOSEMIOTICS.

Luciano Adorini

ENEA-Euratom Immunogenetics Group
Laboratory of Pathology, ENEA C.R.E. Casaccia
C.P. 2400, 00100 Rome A.D., Italy

Peirce, the co-founder with de Saussure of semiotics as an independent discipline, considered it a frame of reference encompassing all other disciplines and thus, although not explicitly mentioned, also Immunology (1). Comforted by this view we can leave to the semioticians the definition of the lower threshold of semiotics (2) and the discussion whether semiotics can be expanded to include immunology.

From a pragmatic point of view it is perhaps more useful to ask whether semiotics can help to understand the immune system. Before trying an holistic approach we could examine the possibility of using semiotic models to analyze simple patterns of immunoregulatory cell interactions.

Redundancy in immunosuppression

In information transmission redundancy is the fraction of the gross information content in a message which can be eliminated without loss of essential information. Redundancy can thus be defined as any deliberate duplication of circuitry or information to decrease the probability of communication failure. In this sense redundancy is an essential component of all communication systems. The concept of redundancy, originally

NATO ASI Series, Vol. H23
The Semiotics of Cellular Communication in the Immune System
Edited by E.E. Sercarz et al.

developed by Shannon (3) has been used by semioticians to understand the rules of message transfer (2, 4) and can be easily applied to the immune system.

For example, the English language is about 50% redundant, in that 50% of the letters in a given message could be eliminated without affecting its intrinsic comprehension. Redundancy, the difference between useful message and transmitted message, is then critical to disambiguate meaning and to provide an appropriate context in which the transmitted message can be placed.

The most important effector systems of the immune response, helper activity, suppression, cytotoxicity and antibody production are redundant. In particular immunosuppression, a critical regulatory function of the immune system, is highly redundant at several levels: these are many different mechanisms able to suppress the immune response, each one involving several cell types communicating via different information systems.

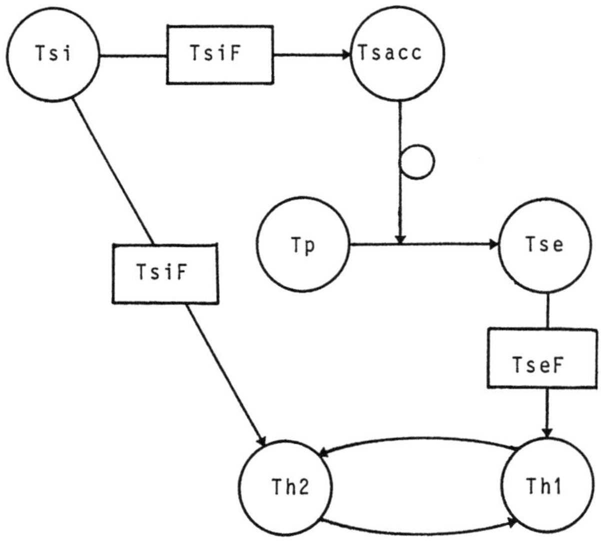

Figure 1. An immunosuppressive circuit as analyzed in the secondary antibody response to a hapten-carrier conjugate. Adapted from Tada (5).

If we consider an antigen-specific suppressor circuit (fig. 1), for example that described by Tada and coworkers (5), it is possible to define at least three levels of redundancy, which could be organized according to their increasing efficiency in information transmission (fig. 2). Obviously, for the purpose of this discussion, it is not relevant how much of the model proposed by Tada is correct.

Repetition of message, the simplest and least efficient form of redundancy, is a common property of cells in the immune system. Any immunologically relevant message like antibodies, interleukins and antigen-specific products is produced in a very large number of copies and mediators of immunosuppressive circuits, in Tada's example T suppressor inducer factor (TsiF) and T suppressor effector factor (TseF), do not represent an exception.

A more efficient mean of message repetition is the amplification of source and receiver, as exemplified in Tada's scheme by T precursor cells (Tp) amplified to effector suppressor T cells (Tse) by T suppressor acceptor cells (Tsacc).

A third level of redundancy is duplication of function, for example the ability of TsiF to directly suppress a set of helper cells (Th2) and to activate at the same time a complex set of amplification loops eventually leading to production of TseF and to suppression of a different set of helper cells (Th1) connected to Th2.

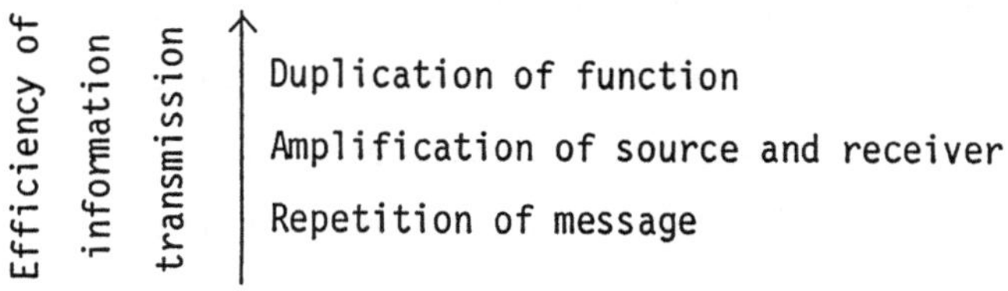

Figure 2.

Teleologically, it is important to appreciate that a vital regulatory system as immunosuppression is highly redundant. Moreover, this immunosemiotic approach indicates that the complexity of immunosuppressive mechanisms is not only due to laboratory artifacts but it reflects the need for the immune system to carry out efficiently a fundamental regulatory function.

Indeed, the problem of efficient and error-free communication is to eliminate from messages superfluous redundancy and to add the type of redundancy able to insure correct message transmission along a disturbed channel. Grammars can be considered as a set of rules to introduce efficient redundancy into a language.

Grammars, machines and immunoregulatory cell interactions

Is has been suggested that knowledge of grammar, hence of language, develops in the child through the interplay of genetically determined principles and a course of experience. This process is known as language learning. In Chomsky's words: "we do not really learn language: rather grammar grows in the mind" (6).

When the immune system develops to its mature form we speak of learning. In both cases, learning a language or learning to discriminate self from non-self, a definite, integrated structure is attained, genetically predetermined and articulated through interaction with the embryological or postnatal environment. Thus, significant components of these cognitive systems are imprinted from a set of restricted alternatives. It is possible to think of language acquisition as a Peirceian abduction (7), a process in which hypotheses are formed according to predetermined rules and are selected according to available evidence. If language learning is considered in these terms then a mind equipped with universal grammar generates alternative grammars that are tested against the data of experience, with the most highly-valued one selected. This metaphor aptly applies not only to the development of the immune system but to the essence of its function, the implementation of selective mechanisms allowing products present in the system prior to the arrival of an external signal to be selected and amplified when an appropriate stimulus is provided.

To assimilate immunology to a cognitive system - and learn something about its nature in this way - we should devote particular attention to domains in which rich and complex structures arise under minimal exposure to the environment, language being a striking example (6). If the immune system is like language then it is properly conceived in terms of grammar-like theories of rules and representations.

Formal language theory deals with mathematical models of languages and the systems used in language generation and processing. The emphasis in language theory is on finite representations of classes of languages based on knowledge of underlying grammatical and syntactic structures. Grammars can thus be described as finite systems that define pattern languages by providing rules for the generation of the language (8).

There is a one-to-one correspondence between the four types of string-generating grammars in Chomsky's hierarchy (9) and four kinds of computing machines, or automata, as recognizers (table 1). In fact, this hierarchy is preserved when the general capabilities of these automata as computing devices are studied. Thus, the Turing machine is, by Church's thesis, equivalent to the most powerful computing device conceivable

(10), whereas the finite state automaton is the least powerful machine of the four.

A grammar with no constraints on the form of its productions (other than the general specification of a finite set of string-rewriting rules) is unrestricted. Grammars, however, fall into a convenient hierarchy as restrictions on productions are imposed. As outlined in table 1 the term "context-sensitive" describes the fact that the nonterminal A can be rewritten as ρ only when A appears in the context of substrings θ and δ. The term "context-free" arises from the fact that the nonterminal A can be rewritten as α regardless of the context in which A appears.

Table 1

Chomsky's hierarchy of string-generating grammars and their correspondence to computing devices as recognizers

Type	Grammar cathegory	Production form	Recognizer
0	unrestricted	no constraints	Turing machine
1	context-sensitive	$\theta A \delta \rightarrow \theta \rho \delta$	Linear-bounded automaton
2	context-free	$A \rightarrow \alpha$	Push-down automaton
3	regular	$A \rightarrow aB$, $A \rightarrow a$	Finite state automaton

A, B, C, non terminals
a, b, c, terminals
α, β, γ, strings of mixed terminals and non terminals

We shall now examine only the simplest language-processing automata, capable of handling languages generated by regular grammars.

A finite automaton is conveniently represented by its state transition diagram, a direct graph with nodes for the states connected by vectors labelled with input symbols causing transitions (fig. 3). By convention all final states appear as double circles and the starting state is indicated by an entering arrow. In the diagram shown in fig. 3 any string of the form a^n causes the automaton to remain in its starting state q_0 but a single occurrence of signal b causes a transition from q_0 to q_1 and q_1 is an accepting state. Hence, strings of the form $a^n b$ are recognized and any other string is not.

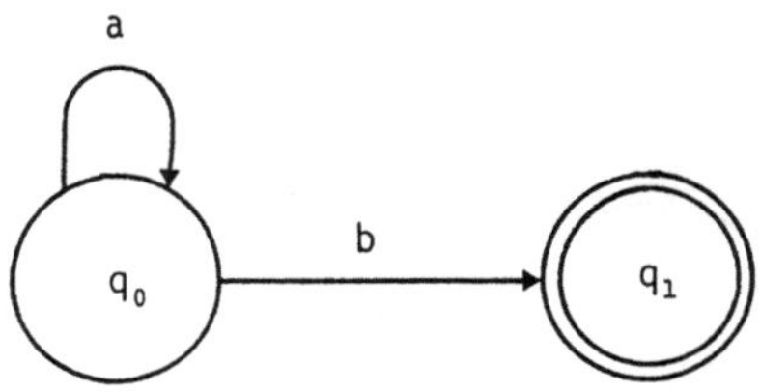

Figure 3. Finite automaton state transition diagram.

The finite automaton whose state transition diagram is given in fig. 4 illustrates the special class of finite automata that are completely specified and deterministic. A completely specified, deterministic finite automaton is one in which, for each state and input combination there is exactly one next state specified by the state transition mapping and the device is never forced to select the next state from among several possibilities. It should be noted that a completely specified finite automaton is a conventional sequential logic network realizable in hardware with flip-flop and combinational logic gates simply assigning output 1 to each final state and output 0 to each nonfinal state.

At this point of the discussion the educated immunologist has already recognized the striking similarity between finite automata and the usual way of describing immunological circuits. However, if expressing immunological circuits with appropriate mathematical notation may be theoretically interesting it is, as of today, pragmatically fruitless. This is because it is not yet possible to provide numerical quantities, for example how many finite states and how many signals do exist in a circuit. Nevertheless, thinking in these terms may help to find solutions to apparently unsolvable problems or at least to rationalize them.

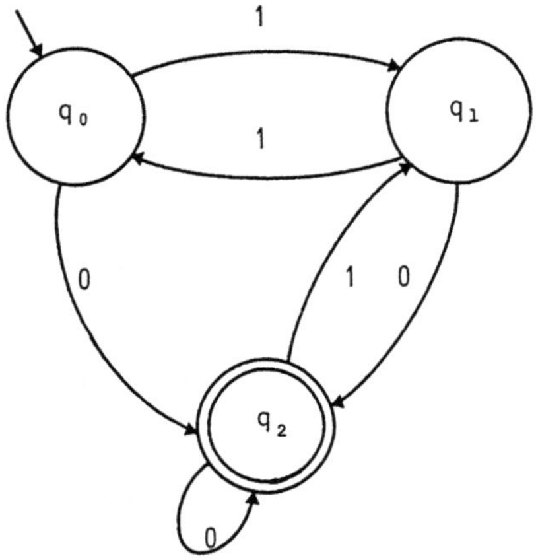

Figure 4. A completely specified and deterministic finite state automaton. The state transistion mapping δ is:

$\delta\ (q_0, 0) = \{q_2\}$ $\delta\ (q_0, 1) = \{q_1\}$

$\delta\ (q_1, 0) = \{q_2\}$ $\delta\ (q_1, 1) = \{q_0\}$

$\delta\ (q_2, 0) = \{q_2\}$ $\delta\ (q_2, 1) = \{q_1\}$

indicating that signal 0 causes all the states to reach the final state q_2 and that signal 1 specifies a new state for each given state.

Conversely, from an immunosemiotic perspective, the discussion on finite state automata to obtain a grammar from a set of patterns could be useful.

The analogy between linguistics and immunology has also been proposed by Jerne (11), who has compared the vast repertoire of the immune system to a lexicon of sentences capable of responding to any sentence expressed by the multitude of antigens the immune system may encounter.

The immune system can also be described by a set of local rules superimposed by a network of connectivity, whose global behaviour generates a cognitive system (F. Varela, this volume). The local rules could be exemplified by finite state automata, real cellular automata following neighborhood rules specified by a regular grammar-type set of rules and representations. The accepted general theory to explain the global behaviour of the immune system is the network theory, a viable approach to immunosemiotics. A reductionist approach - cellular automata or molecular biology - will describe only single components and possibly their mechanism of action but it will never permit understanding of the system. The global approach requires to invent a biological information processing system which could behave like the immune system. The system would then accomodate all the immune components and explain their behaviour. In other words, when the immune network theory will generate a cognitive system, able to meaningfully discriminate relevant from non relevant and self from non-self, an immunosemiotic approach will be available.

There is a long way to immunosemiotics.

References

1. Lieb, I.C. (Ed.) "Charles S. Peirce's letters to Lady Welby", Witlock's, New Haven 1953.
2. Eco, U. "Trattato di Semiotica Generale", Bompiani, Milano 1975.
3. Shannon, C.E., A mathematical theory for communication. Bell System Technical Journal 27: 379, 1948.
4. Eco, U. "Semiotica e Filosofia del Linguaggio", Einaudi, Torino 1984.
5. Tada, T., Help, suppression and specific factors. In: "Fundamental Immunology", Paul W.E. (Ed.), Raven Press, New York 1984.
6. Chomsky, N., Rules and representations. The Behavioural and Brain Sciences 3: 1, 1980.

7. Peirce, C.S., Terms, propositions and arguments. In: "Collected Papers", Harvard University Press, Cambridge 1931-58.
8. Gonzales, R.C. & Thomason, M.G. "Syntactic Pattern Recognition. An Introduction", Addison-Wesley, Reading 1978.
9. Chomsky, N., On certain formal properties of grammars. Inform. Contr. 2: 137, 1959.
10. Davis, M. "The Undecidable: Basic Papers in Undecidable Propositions, Unsolvable Problems and Computable Functions", Raven Press, New York 1968.
11. Jerne, N.K., The generative grammar of the immune system. EMBO J. 4: 847, 1985.

IS THERE A UNIQUE LANGUAGE IN THE IMMUNOREGULATORY SYSTEM?

MASARU TANIGUCHI

Department of Immunology, School of Medicine, Chiba University, Chiba, Japan.

NATO ASI Series, Vol. H23
The Semiotics of Cellular Communication in the Immune System
Edited by E. E. Sercarz et al.

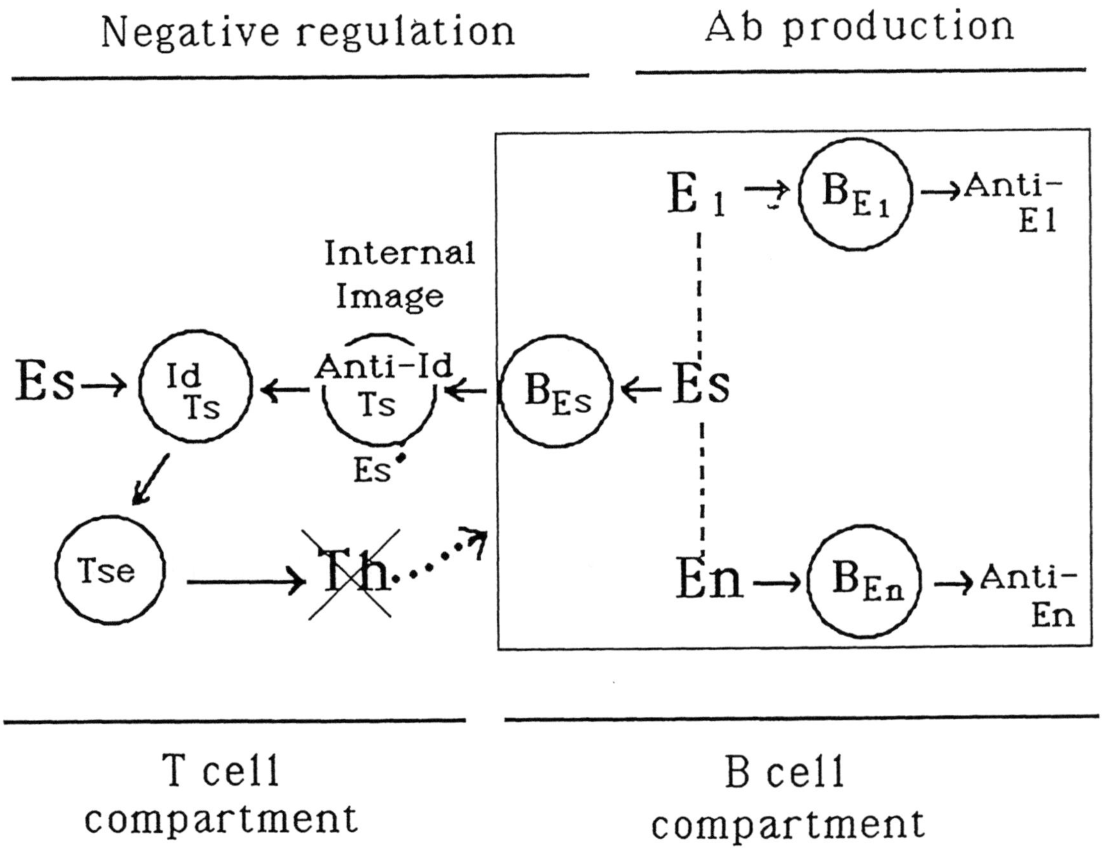

Figure 1. Possible cell interactions in antibody formation.

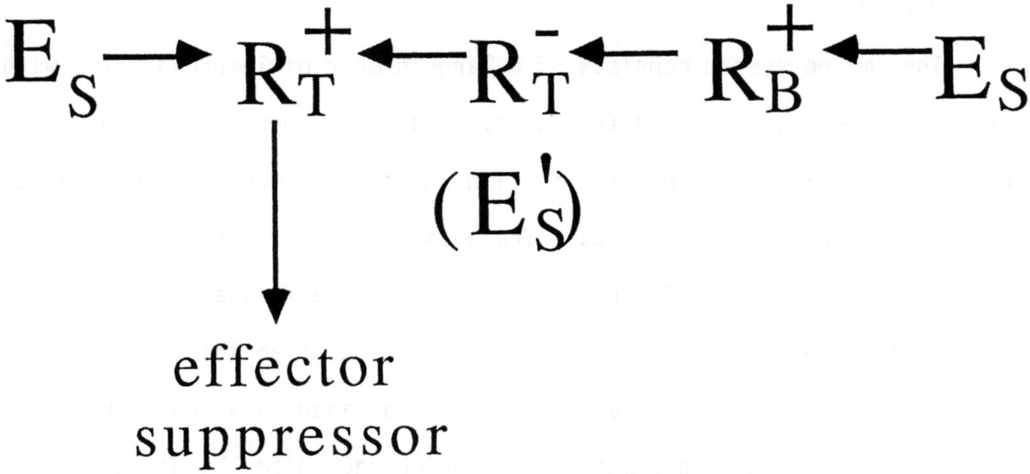

Figure 2. Symmetrical network in the induction phase of regulatory cell interactions.

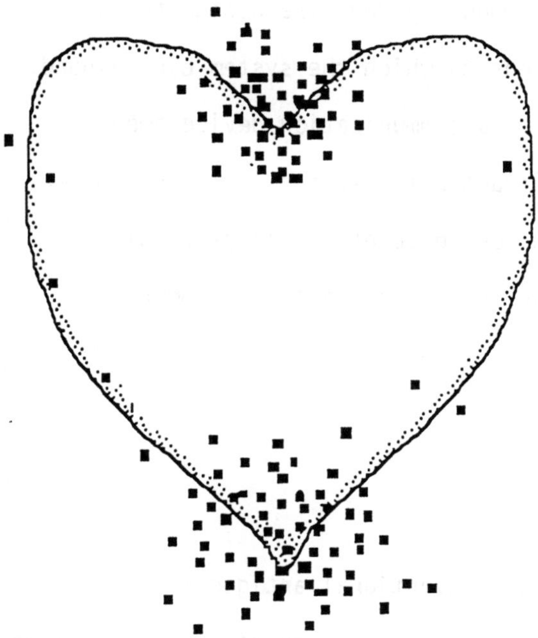

Figure 3. Viewpoints when look at the shape of the heart.

INTRODUCTION

The immune system consists of a large number of lymphoid cells with different specificities and functions. Each cell has an antigen receptor with a unique specificity with which the lymphocyte responds to a particular antigenic stimulus. The antigen binding site of a particular antibody or of a receptor molecule itself determines to specificity, being accompanied by the expression of an idiotype (Id). Idiotypic determinants generally possess an antigenicity, and can induce antibodies (anti-idiotypic antibodies) which recognize the variable region of the original antibody. Thus, the immune system consists of a complex network in which antibody idiotypes, and anti-idiotypic antibodies function as both <u>antigens</u> and antibodies (1). The idiotypic determinants, <u>antigens</u> inside the immune system, are designated as "internal images " for outer antigens to which the system can respond. This internal image is often used as a communication device among lymphocytes to convey particular structural messages. In this paper, the biological significance and unique features of the particular messages (internal image) appearing in the regulation of immune responses are discussed.

FACTS

I. When animals are immunized with conventional antigen with multi-antigenic epitopes, like keyhole limpet hemocyanin (KLH), various B cell clones are stimulated and produce antibodies with different KLH specificities. No predominant idiotypes (major idiotype) on anti-KLH antibodies have so far been detected at the antibody level. This indicates that there is no specific epitope which B cells preferentially recognize and that these KLH epitopes are common languages used in B cells. B cells can thus produce antibodies to every epitope. On the other hand, T cells seem to recognize very restricted epitopes only. This was originally demonstrated by Sercarz et al. (2) and others (3). For example, Sercarz et al. (2) showed that helper or suppressor T cells which regulate immune responses positively or negatively recognize certain epitopes on different portions of the same antigen, denoted as helper and suppressor epitopes, respectively. In their experimental system using hen egg lysozyme (HEL), a peptide fragment 1-17 of HEL is the immunodominant region for suppressor T cells, whereas helper T cells preferentially recognize the fragments 13-35 and 74-96 of the same molecule. Therefore, these epitopes are regarded to be specific targets which helper or suppressor T cells tend to recognize and would function as communication vehicles for either helper or suppressor T cells to be activated. Thus, functionally distinct lymphoid cell populations recognize the particular epitopes on a single molecule. This indicates that lymphocyte subsets use their own specialized languages according to the regulatory principle of their system. This also implies that helper and suppressor T cells and B cells possess their own individual idiotype network systems, a part of which should be overlapping.

II. The complex cell to cell interactions in immune regulation can be analysed by using the somatic cell hybrid technique. Since Koehler and Milstein (4) developed this technique for the isolation of monoclonal cell lines with predefined specificity, it has become possible to isolate T cell lines and test the network hypothesis in regulatory T cell interactions.

We have established two sets of KLH-specific suppressor T cell hybridomas (34S-704 and 34S-281) made by fusion of the BW5147 thymoma and C57BL/6 ($H-2^b$) suppressor T cells. 34S-704 is idiotypic and binds the antigen KLH, whereas 34S-281 is anti-idiotypic and does not bind KLH. These hybridomas are complementary to each other and both exert KLH-specific and $H-2^b$ restricted suppressor activity. Interestingly, the anti-idiotypic hybridoma expresses the structure mimicking the suppressor epitope of KLH presented on antigen-presenting cells (APC) (5). This structure is called the "internal image".

Moreover, only by this internal image, suppressor T cells and B cells recognizing suppressor epitopes can communicate with each other. In fact, we successfully established monoclonal anti-KLH antibodies reactive with anti-idiotypic suppressor T cell hybridoma (345-281) (6). The anti-KLH antibodies reacted with the internal image on the anti-idiotypic suppressor T cells and stimulated these T cells to be active. The activated anti-idiotypic T cells secrete functional molecules which in turn stimulate idiotypic suppressor T cells, transmitting signals to the effector cells. These effector T cells prohibit the antibody production of B cells by suppressing the helper T cell function.

III. The structure of "the internal image of suppressor epitope on KLH " is not the same as the antigen KLH fragment itself but is that modified with the self molecule. This idea was obtained from the following experiments (7). Fully allogeneic, X-irradiated bone marrow chimera mice (C3H—— B6 mice: C3H bone marrow cells were transferred into X-irradiated host C57BL/6 mice) were immunized with the antigen (KLH) and used as responding cells for antibody formation. When the host type APC were added back to the chimeric mice at the time of immunization, suppressor T cells acquired the host type (C57BL/6:H-2^b) restriction specificity. As suppressor T cells originated from C3H (H-2^K) mice, this result indicates that C3H (H-2^k) suppressor T cells with KLH and H-2^b restriction specificity were generated in the presence of C57BL/6 (H-2^b) APC.

Moreover, the immune responses of the C3H——B6 chimera mice receiving host APC were suppressed by monoclonal anti-KLH antibody that was only effective in the H-2^b mice. In the chimera conditions, C3H derived anti-idiotypic suppressor T cells express the internal image which mimics the "KLH modified with H-2^b self like structure". Therefore, the structure of the internal image resembles the antigen plus the self molecule. Based on these data, the possible scheme of cell interactions in the immune responses is illustrated in Figure 1.

SYMMETRICAL NETWORK IN THE REGULATORY CELL INTERACTIONS

The initial regulatory cell interactions described above can be simplified by the symmetrical network system (Figure 2). Two classes of lymphocyte receptors are involved in this reaction. The first set is the plus or antigen-binding set, denoted R_T^+ and R_B^+ for T cell and B cell, respectively. The second class is the minus or anti-idiotypic set, R_T^-, harboring receptors specific for R^+. The ligand Es(suppressor epitope) is a unique epitope for suppressor cells. In the suppressor pathway, R_T^- expresses the internal image of the suppressor epitope. Therefore, R_T^- mimics Es, denoted Es'. Signals from Es or Es' always construct positive stimuli, and the direction of the signs is integrated into R_T^+ cells, because only R_T^+ cells transmit signals to the effector suppressor phase.

Based on this notion, the regulatory cell interactions can be explained as follows.

1). APC presents the "special sign" (Es) to the suppressor cell family.

2). Idiotypic suppressor T cells (bearing R_T^+ receptors) work as transmitters which receive signals from antigen (Es) and transfer them to the cells in the effector phase.

3). The transmitter also receives signals from the anti-idiotypic suppressor T cell which is an "icon-bearer" expressing the internal image (Es). The "icon-bearer" (with R_T^- receptor) itself also works as an interpreter which receives signals from some other lymphoid populations (B cells recognizing suppressor epitope:B_{Es}).

4). The structure of "icon" (R_T^-) is composed of antigen plus self elements, and mimics the sign for suppressor T cells (Es) which is expressed on antigen-presenting cells.

WHY DOES THE T CELL RECOGNIZE ONLY RESTRICTED ANTIGENIC EPITOPES?

In general, B cells recognize a variety of epitopes on single molecule, whereas T cells see only a particular epitope which characterizes the immunogenicity of the antigen. The mode of this T cell antigen recognition is similar to that of the central nervous

system when recognizing the object. For example, if we casually look at the shape of the heart, the major viewpoints are those of the V-shaped parts of the heart but not of other parts(Figure 3)(8).
The most typical structure of the heart for visual sensory neurons is thus the V-shaped configuration. In a similar manner, T cells see only the restricted part of the antigen. T cells and neurons understand the whole structure of the object by recognizing the most characteristic parts. The mode of this recognition by T cells is quite economical in terms of regulatory cell interactions when compared with that by B cells. Due to the uniqueness of the T cell recognition, the T cell repertoire is possibly rather limited compared to that of B cells.

Jerne's original network hypothesis (1) is advantageous for explaining the mechanisms of clonal expansion and maintenance of the B cell repertoire. However, if T cells are destined to see every epitope which B cells see, and if T cells have a huge idiotype network corresponding to all antigenic epitopes, enormous negative energy would be necessary in order to accomplish the effective regulation of immune responses to various epitopes. Therefore, it is assumed that a more economical mechanism should be operative, particularly in the regulatory interactions.

B cell responses are nonspecifically controlled by helper T cells. Similarly, the effector suppressor T cells nonspecifically regulate helper T cells, resulting in the control of B cells. Therefore, one suppressor T cell clone recognizing one epitope is enough to control whole helper T cell functions and would subsequently shut off the responses of various B cell clones. This must be true in physiologic immune responses, as a monoclonal KLH-specific suppressor T cell clone or hybridoma having only one specificity can suppress the total B cell response to various antigens in an antigen-specific manner (7). Thus, the limited T cell repertoire is advantageous for the effective regulatory system.

WHAT DOES THE INTERNAL IMAGE DO IN REGULATORY CELL INTERACTIONS?

As mentioned above, T cells possess a unique language system. The internal image expressed on suppressor T cells is used as a language only for the suppressor cell family. In particular, this language is essential for functional communication between suppressor T cells and B cells, as well as between idiotypic and anti-idiotypic T cells. We call this communication device the regulatory internal image. Among antibodies against external antigens, only the antibody reactive to the regulatory internal image can interact with and activate suppressor T cells bearing icons of suppressor epitopes (internal image). Thus, the internal image functions not only as an icon of a suppressor epitope, but also as the transmitter or interpreter which receives signals from antibody and transfers them into the cells in the effector pathway. It is thus evident that this regulatory internal image has the biologically incisive function of maintaining the suppressive state via feedback mechanisms even after antigen elimination from the immune system has been completed.

Other functional meanings of the internal image may also have to be considered. The internal image should work as the presenter of epitopes to specifically activate the suppressor family. This particular function may be important in determining the repertoire of regulatory T cells, and thus acts as the dictator of suppressor T cells. In fact, our chimera experiments (7) have demonstrated that APC are needed for the clonal expansion of suppressor T cells, and for the determination of the antigen and genetic restriction specificities of suppressor T cells. Moreover, monoclonal anti-idiotypic suppressor T cell hybridoma expressing the internal image of KLH induced KLH-specific suppression of antibody responses. Therefore, it appears that the internal image as well as APC, like the thymus, can determine the repertoire of suppressor T cells.

As the regulatory internal image possesses the functional structure mimicking the "antigen modified self molecule" and is used as a communication device, suppressor T cells have to recognize the genetic element in their interactions. Thus, the "genetic restriction" is always maintained. In this regard, the element involved in the genetic restriction in cellular interactions is similar to the synaptic element seen beween neurons in the central nervous system. With this element, suppressor cells are able to find partners and "communicate with" them.

Taken collectively, the internal image displayed on suppressor T cell receptors is a communication device for regulatory cell interactions, and may contribute to the selection and maintenance of the suppressor T cell repertoire. B cells recognizing the internal image may also influence the repertoire of suppressor T cells.

REFERENCES

1. Jerne, N. K.(1974). Towards a network theory of the immune response. Ann. Immunol. (Paris) 125C:373.

2. Sercarz, E. E., Yowell, R. L., Turkin, D., Miller, A., Araneo, B. A., and Adorini, L. (1978). Differential functional specificity repertories for suppressor and helper T cells. Immunol.Rev.39:108.

3. Benjamin, D. C., Berzofsky, J. A., East, I. J. et al. (1984). The antigenic structure of proteins: A reappraisal. Ann. Rev. Immunol. 2:67.

4. Koehler, G. and Milstein C.(1975). Continuous cultures of fused cells secreting antibody of predefined specificity. Nature 256:495.

5. Takei, I., Sumida, T., and Taniguchi, M. (1983). Acceptor-suppressor T cell hybridoma with receptor recongnizing antigen-specific suppresssor factor. J. Exp. Med. 158:1912.

6. Sumida, T. and Taniguchi, M. (1985). Novel mechanisms of specific suppression of anti-hapten antibody response mediated by monoclonal anti-carrier antibody. J. Immunol. 134:3675.

7. Sumida, T., Sado, T., Kojima, M., Ono, K., Kamisaku, H. and Taniguchi, M. (1985). The I-J as an idiotype of recognition component of antigen-specific suppressor T cell factor. Nature 316:738.

8. Zusne, L., and Michels, K.M.(Purdue Univ.Press)(1964). Nonrepresentational shapes and eye movements. Percept. Motor Skills, 18(1),11-20.

ACKNOWLEDGMENTS

This work was supported in part by grants from the Ministry of Education, Culture and Science, Japan, and from the Mochida Memorial Foundation, and the Health Science Foundation. I am grateful to Drs. Takeshi Tokuhisa, Masatoshi Tagawa and Professor Tomio Tada, Tokyo University for their generous suggestions and discussions. I also wish to express my appreciation to Ms. Rieko Shimomura and Ms. Sachie Kurata for typing this manuscript.

RECOGNITION OF SYNGENEIC FORMS OF L-CHAIN V-REGION PEPTIDES BY T HELPER CELLS SUGGESTS THAT B CELL ANTIGEN RECEPTORS ARE PROCESSED AND REUTILIZED IN COGNATE T-B INTERACTION

Kristian Hannestad, Gunn Kristoffersen and Jean Paul Briand
Intitute of Medical Biology, University of Tromsø, Tromsø and
Institut de Biologie Moleculaire et Cellulaire, Strasbourg

1. The myeloma protein 315 system

Idiotope-specific resistance to a malignant B cell line. We have worked with the monoclonal Ab produced by plasmacytoma 315, a high-rate Ab-producing malignant B-cell line originating in the inbred mouse strain BALB/c. We wondered whether T helper cells of BALB/c mice were capable of recognizing small peptides from the third hypervariable region (HV3) of the 315 L-chain. One reason for asking this question was that mice immunized with M315 develop resistance to challenge with a small number of 315 plasmacytoma cells that kill other mice immunized with another monoclonal Ab with the same H-chain C-region. Furthermore, the M315-immunized mice were killed by a small number of cells from another plasmacytoma that produced Ab with the same C_H region as M315 (1). This indicated that the immune-mediated resistance to the tumor was directed towards V-regions, i.e. idiotopes, bound on the surface of the tumor cells (2). Idiotopes may therefore function as tumor-specific transplantation antigens, and we wanted to know which part of the Ab molecules was being recognized. In addition, we hoped to learn something more general about how T cell clones within an immune system recognize B cell clones, because of the possibility of network interactions (3).

T cell defined diotopes of myeloma protein 315: previous observations. From previous studies by Sirisinha and Eisen (4) it was known that M315 elicits an Ab response to idiotopes, and that these Ab define idiotopes require associated H- and L-chains for their expression, i.e. the Abs did not bind the

NATO ASI Series, Vol. H23
The Semiotics of Cellular Communication in the Immune System
Edited by E. E. Sercarz et al.

isolated chains. The results of studies performed during the last few years in our laboratory indicated that T helper cells also recognize idiotopes of M315 but in contrast to the serologically defined idiotopes, T cell defined idiotopes were expressed by isolated V-regions. Furthermore, we had found that the histocompatibility molecule (Ia) variant of the mouse engaged in an immune response to M315 determined which of its two V-regions, Vλ2 or V_H, was recognized by T helper cells. Thus, the Ia variant of BALB/c mice (H-2^d) directed recognition to the V λ2 region, whereas the H-2^s and H-2^k variants were associated with recognition of the V_H region (5, 6).

2. The simplicity of lambda V genes

Our studies have lately focused on the mouse T-cell defined idiotopes expressed by the lambda L-chain of plasmacytoma MOPC315. The gene family encoding lambda L-chains of inbred mice is unique in that in contains only two germline V-gene segments encoding amino acids 1-97 and three functional J-gene segments encoding amino acids 99-109 (see ref. 7 for review). By contrast, there exists at least 200 V-kappa gene segments and from 100 to more than 1000 V_H-gene segments. The two V-gene segments, Vλ1 and Vλ2, share extensive DNA sequence similarity and their corresponding amino acid sequences differ at only 7-8 positions. The 315 V λ2 gene segment has undergone somatic mutations and encodes a polypeptide which differs at four positions (38, 94, 95, 96) from the germ line Vλ2 gene amino acid sequence. This homology between lambda chains offers special opportunity to study the contribution of individual amino acids to T helper cell defined idiotopes, i.e. structural correlates of idiotopes.

3. Cognate T-B interaction and T cell help

Activation of resting B lymphocytes to divide and develop into high rate Ab-forming cells involves at least two triggering events: 1) binding of antigen by the paratopes of the Ab embedded in the B-cell surface, and 2) T-cell help. T-cell help for B-cell activation appears to be mediated by both

Table 1

Amino acids at the Vλ2-Jλ2 junction are critical for T-cell defined idiotopes of λ2 L-chains

Antigen used for: T helper priming	Antigen used for: Boost of recipients	Distinct amino acid of boost antigen: 98	Distinct amino acid of boost antigen: 99	Secondary anti-NIP Ab response of recipients (μg/ml)
H_2O	germline λ2	Y	V	<4
germline λ2	germline λ2	Y	V	57
germline λ2	T952 λ2	F	I	<4

See Section 4 for experimental design. Except for the indicated positions the amino acid sequences of germline and T952λ2 chains are presumably identical (7). The Ab concentration was determined by solid phase ELISA technic.

Table 2

Contribution of mutated amino acid residues at positions 94, 95, 95 to T helper cell defined idiotopes of 315 λ2 chain

Priming antigen for T helper cell	Amino acid position of priming peptide: 94	95	96	Secondary anti-NIP Ab response of recipients boosted with NIP-315 λ2 (μg/ml)
H_2O				<8
Peptide				
1-117	F	R	N	125
91-108	-	-	-	95
91-108	Y	-	-	40
91-108	-	S	-	<8
91-108	-	-	T	10

See section 4 for experimental design.
Peptide 1-117 was a cleavage peptide which contains the 315 (V+J) region plus 8 amino acids of the Cλ2-region. Peptides 91-108 were synthetic. FRN are the mutated amino acids at the indicated positions of 315 λ2 chain, and YST are the corresponding amino acids of the germline gene encoded λ2 sequence.

secretion of non-antigen specific soluble chemical signals, lymphokines, and direct T-B cell interaction. Direct T-B cell interaction involves binding of stimulation complexes for T-cells (Ia plus processed antigen) on the surface of the B-cell by antigen receptors on the surface of the T-cells. This binding is called cognate T-B interaction. We postulated that V-regions of Abs can form stimulation complexes with Ia. Thus, if the processed antigen is derived from V-regions of a monoclonal Ab, then the recognition of idiotopes by T-cells can be studied by the cognate interaction of idiotope-specific T-cells with B-cells presenting V-region peptides from the monoclonal Ab.

4. Method to study recognition of idiotopes by T helper cells

The recognition of 315 λ2 L-chains by T helper cells was studied by taking advantage of their function in cognate T-B interaction, an experimental approach pioneered by Mitchison. To prime T helper cells, BALB/c mice were immunized in the footpads with 1) a fragment of 315 λ2 L-chain that contained the entire (V+J) region, or 2) synthetic peptides spanning the 315 λ2 amino acid region 91-108, which includes HV3. The antigens were emulsified in complete Freund's adjuvant. After 10 days the draining popliteal lymphnodes were harvested, and single cells were transfused into X-irradiated (500 rad) BALB/c recipients together with spleen cells containing B lymphocytes primed with a small chemically defined group, nitro-iodo-phenol (NIP). The recipients were then injected with NIP coupled to 315 λ2 L-chains (NIP-315λ2) to "boost" the primed cells. If T helper cells primed with 315 Vλ2 or synthetic peptide engage in cognate interaction with primed B cells that have bound NIP-315 λ2 (the boost antigen), a brisk anti-NIP Ab response in the recipient is detectable 9 days later. A minimal anti-NIP response would indicate lack of cognate T-B interaction.

5. Results

A more complete presentation of these results has been published (8).

T helper cells recognize the Vλ2-Jλ2 junction. To explore the discriminating ability of T helper cells for variant λ2 L-chains, we took advantage of the availability of T952 and 5-7 chains which only differ at position 98 and 99 at the Vλ2-Jλ2 junction (Table 1). T helper cells primed with the 5-7 λ2 chain responded well to a boost with the immunogen as shown by a strong anti-NIP Ab response but they did not respond to a challenge with T952 λ2 chain (Table 1). This indicated that T helper cells can discriminate between two Ig L-chains containing 215 amino acids of which 213 are identical.

T helper cells respond to short synthetic peptides spanning HV3 of 315 λ2. T helper cells primed with synthetic peptide spanning region 91-108 (HV3 and J region) of the 315 λ2 chain responded to a boost with NIP-315 λ2. This demonstrated that an 18 amino acid peptide was sufficient to prime the T helper cells for a response to the complete L-chain. Lymphocytes from animals primed with shorter peptides (88-99 and 94-108) of 315 λ2 did not respond to the boost antigens (data not shown).

Two of three mutated amino acids are critical for T helper cell defined idiotopes. The 315 λ2 chain contains four somatic amino acid replacement (mutations) in position 94, 95 and 96. To determine whether the mutated amino acids in the HV3 (positions 94, 95, 96) were essential for the T helper cell defined idiotopes, three peptide analogs were synthesized that represented amino acid region 91-108 of 315 λ2 except at one of these three positions which was substituted with the corresponding germline gene encoded amino acid. Four out of six recipients of lymphocytes primed with the Thr^{94}-containing peptide responded well (40 μg anti-NIP Ab/ml) to a challenge with NIP-315 λ2 (Table 2). By contrast, T helper cells from animals primed with the Ser^{95}- or Thr^{96}-containing peptide did not support augmented anti-NIP Ab responses following a boost with NIP-315 λ2.

6. Discussion

The data establish that T helper cells can recognize a small peptide of 18 amino acids spanning HV3 and the Jλ2-region of a syngeneic L-chain, and that the T-cells discriminate between single amino acid substitutions at certain positions of this peptide. Furthermore, the cells can distinguish between two λ2 L-chains that only differ by two amino acids at the Vλ2/Jλ2 junction. Thus the fine specificity of the T helper cells for syngeneic L-chains is striking. The simplest explanation for these results is that these T helper cells recognize peptides processed by antigen-presenting cells. According to this model and assuming that the anti-NIP Ab responses represent cognate T-B interaction, B-cells that have bound NIP-315 λ2 L-chains (the boost antigen) are likely to have processed this chain to peptides resembling the 91-108 peptide. Previous results from our laboratory showing that the specificities of the T-cells in this system is strongly influenced by immune response (Ir) genes linked to the major histocompatibility complex further strongly indicate that the processed peptides bind to Ia molecules to form stimulation complexes for T-cells. Other peptide regions of the λ2 chain may be less immunogenic because they bind less firmly to the $H-2^d$ variant of Ia molecules or, bearing in mind that the antigen is a syngeneic L-chain, because self-tolerance limits the available repertoire of T-cells for these regions. In this context it may be highly significant that the mutated amino acids of HV3 of 315 λ2 were critical for the specificity of the T-cells, suggesting that self-tolerance did not exist for the mutated 315 λ2 variant. Mutations located in peptide regions of Igs that form efficient complexes with Ia molecules may therefore greatly enhance immunogenicity of B cell antigen receptors. Thus, B-cell clones bearing mutated antigen receptors may, provided they are able to process these receptors, come under the influence of mutation-specific regulatory T-cells. These T helper cells would selectively interact with and help the mutated clones and thus contribute to anamnestic responses and affinity maturation of Abs.

This idea envisages that two distinct categories of stimulation complexes can form on the surface of a B lymphocyte that has bound antigen: one category consists of processed peptide from the conventional bound antigen, the other contains peptides derived from the antigen receptor of that B-cell. If the conventional (nominal) antigen is weakly immunogenic, the contribution of stimulation complexes containing peptides derived from the clonal antigen receptor could conceivably play a dominating role in cognate T-B interaction. With strongly immunogenic nominal antigens this effect may be hard to discern. We have suggested a similar hypothesis before (9), and recently others have also arrived at related models (10-12).

Processing of antigen recognition molecules on B cells could be triggered by antigen binding. Thus, the receptor first captures antigen and is then subsequently internalized and reutilized in processed form by engaging in cognate T-B interaction. Alternatively, B cells could process some of the Ig chains they synthesize before the chains are assembled intracellularly to complete Ig molecules. The presentation of processed clonotypic antigen receptors may be very efficient relative to conventional protein antigens due to high concentration and homogeneity of receptor available to the B cell which synthesizes that receptor.

However, some B cell clones would bear receptors unsuitable for participation in cognate T-B interaction due to Ir-gene defect, as is the case for both V_H and Vλ2 domains of M315 in $H\text{-}2^b$ haplotype mice which are non-responders to these domains (6). If Fab regions are relatively resistant to processing this could represent another restriction; possibly, IgD forms of the receptor may be more degradable than IgM (13). Thirdly, deletion or suppression of T cell clones due to immunological tolerance could restrict the T cell repertoire for those germ-line gene encoded V-domains that are present in tolerogenic concentrations. In this context it is interesting that the germ-line gene encoded 5-7 λ2-chain was immunogenic. This chain has Tyr in position 98 at the site of Vλ2-Jλ2 recombination. The other alternative of flexible Vλ2-Jλ2 joining is Phe^{98} (7). Possibly, Tyr^{98} is a rare variant (relative to Phe) for which tolerance has not developed; otherwise one would expect that the concentration of 2-bearing Ig, which is approximately 80 µg/ml in BALB/c mice (14), should have induced non-responsiveness to non-mutated 2-chains like 5-7.

If Id-recognition by conventional T helper cells requires processing, which the present results strongly suggest, it follows that the vast structural repertoire of B cell antigen receptors contains two categories of internal images of foreign or of auto-antigens: (a) the classical serologically defined internal image which is likely to be destroyed by processing, and (b) internal images defined by conventional T cells which require denaturation and formation of stimulation complexes with Ia before they can address T helper cells. It is possible that internal images of the second category could mimic auto-antigens and sensitize T cells to self.

Concluding remarks

Conventional T helper cells probably recognize syngeneic forms of Ig V-regions (idiotopes) that have been processed to small peptides and presented in context of Ia. Two of three somatic amino acid replacements (mutations) in HV3 of a λ2 L-chain were shown to be critical for T helper cell defined idiotopes. The data open the possibility that

1) the clonotypic antigen recognition molecules of B cells can be reutilized in cognate T-B interaction, after they have been processed,
2) B cell clones bearing mutated or rare variants of antigen receptors may be selected by interaction with idiotope-specific T helper cells:
3) a B cell may bear two categories of internal images: (a) the classical serologically defined, and (b) T cell defined internal images which consist of peptides from the antigen receptor in complex with Ia.

Literature Cited

1. Lynch, R.G., Graff, R.T., Sirisinha, S., Simms, E. and Eisen, H.N., Myeloma proteins as tumor-specific transplantation antigens. Proc. Natl. Acad. Sci. USA 69, 1540-1544 (1972).

2. Hannestad, K., Kao, M.S. and Eisen, H.N., Cell-bound myeloma proteins on the surface of myeloma cells: Potential targets for the immune system. Proc. Natl. Acad. Sci. USA 69, 2295-2299 (1972).

3. Jerne, N.K., Towards a network theory of the immune system. Ann. Immunol. (paris) 125c, 373-389 (1974).

4. Sirisinha, S. and Eisen, H.N., Auto-immune like antibodies to the ligand binding sites of myeloma proteins. Proc. Natl. Acad. Sci. USA 68, 3130-3135 (1971).

5. Jørgensen, T. and Hannestad, K., Helper T cell recognition of the variable domains of a mouse myeloma protein (315). Effect of the major histocompatibility complex and domain conformation. J. Exp. Med. 155, 1587-1596 (1982).

6. Jørgensen, T., Bogen, B. and Hannestad, K., H-2 linked Ir-genes have a striking influence on the immunogenicity of myeloma protein 315 for T helper cells. Scand. J. Immunol. 21, 183-187 (1985).

7. Eisen, H.N. and Reilly, E.B., Lambda chains and genes in inbred mice. Ann. Rev. Immunol. 3, 337-365 (1985).

8. Hannestad, K., Kristoffersen, G. and Briand,J.P., The T lymphocyte response to syngeneic λ2 light chain idiotopes. Significance of individual amino acids revealed by variant λ2 chains and idiotope-mimicking chemically synthesized peptides. Eur. J. Immunol. 16, 889-893 (1986).

9. Jørgensen, T., Bogen, B. and Hannestad, K., Recognition of variable (V) domains of myeloma protein 315 by B- and T-lymphocytes. In Janeway, C., Sercarz, E. and Wigzell, H. (Eds.) Immunoglobulin Idiotypes. ICN-UCLA Symposia on Molecular and Cellular Biology, vol XX, Academic Press, 1981, p. 573.

10. Leserman, L., The introversion of the immune response: a hypothesis for T-B interaction. Immunol. Today 6, 352-355 (1985).

11. Kawahara, D.J., Marrack, P. and Kappler, J.W., H-2-linked Ir gene control of V_H determinant(s)-specific helper T cells. J. Mol. Cell. Immunol. 2, 255-264 (1986).

12. Kourilsky, P. and Claverie, J.-M., The peptidic self model: a hypothesis on the molecular nature of the immunological self. Ann. Immunol. (Paris) 137D, 3-21 (1986).

13. Bourgois, A., Abney, E.R. and Parkhouse, R.M.E., Mouse immunoglobulin receptors on lymphocytes: identification of IgM and IgD molecules by tryptic cleavage and a postulated role for cell surface IgD.

14. Cotner, T. and Eisen, H.N., The natural abundance of λ2-light chains in inbred mice. J. Exp. Med. 148, 1388-1399 (1978).

MORE!

REGULATORY T CELL FUNCTION, ANTIGEN RECOGNITION, AND AN IMMUNOLOGICAL DECISION POINT

DOUGLAS R. GREEN
Department of Immunology
University of Alberta
Edmonton, Alberta, CANADA T6G 2H7

INTRODUCTION

Since their discovery, thymus dependent (T) lymphocytes have held the promise of a key to the fundamental mystery of the immune system: the discrimination of self from nonself. There are two reasons for this. The first is that helper T cells, which are required for the vast majority of immune responses, only respond to their antigens in the presence of products of the major histocompatibility complex (MHC) (1). Thus, they respond to a combination of nonself (antigen) and a form or marker of self (class II MHC molecules). The second reason is that another type of T cell, the suppressor T cell, is capable of inhibiting immune responses to specific antigens. Thus, one can imagine that these cells normally prevent unwanted responses to self antigens while allowing responses to non-self antigens. In fact, examples of such control have been described (2,3). Unlike helper T cells, however, the antigen-specific products of suppressor T cells respond to antigen in the absence of MHC molecules. One therefore has the feeling that if we could fully understand the basis for each type of response to antigen (help vs suppression) we would reveal the hidden mechanisms of self-nonself discrimination.

This is, of course, a reductionist viewpoint, and its partial failure is already apparent. Tremendous progress has been made in the investigation of the T cell receptor of helper T cells which is responsible for the response to foreign antigen plus self MHC molecules (4). Unfortunately, however, we seem to be no closer to understanding how the immune system "knows" what it is responding to: self or nonself. The problem is worse for suppressor T cells;

NATO ASI Series, Vol. H23
The Semiotics of Cellular Communication in the Immune System
Edited by E. E. Sercarz et al.

not only do we not understand how such cells respond to self vs nonself, but we also have no clear idea of the molecular basis for their response to antigen.

The reader must be warned at the outset that we do not hope to solve this fundamental mystery in the next few pages. We do hope, however, to present a way to study certain interactions in the immune system which may shed new light on the problem. Our approach, while unabashedly reductionist, owes its allegience to semiotics in its focus on decision making processes which impart information to the system. We address one process in particular, the decision between help and suppression, and in so doing find a new relationship between immunoregulatory function and the way in which antigen is recognized (we consider the term, "recognize" in more detail below). We begin by examining a more "traditional"* view of immune recognition and regulation, followed by that which we propose.

A "TRADITIONAL" VIEW OF HELP AND SUPPRESSION (Figure 1)

As discussed by several authors in this volume, helper T cells respond to specific antigen only when it is presented together with specific MHC molecules. When activated in such a way, helper T cells release antigen nonspecific lymphokines which induce other cells, such as B cells and macrophages, to perform immune effector functions. Helper T cells, in the mouse, generally bear cell surface molecules detected by antibodies specific for L3T4 and Ly-1 (5).

Suppression of the immune response is slightly more complex. T suppressor inducer cells (Tsi) release antigen specific molecules which induce T suppressor effector cells to release a second set of antigen specific molecules responsible for inhibition of immunity (6,7). The antigen specific molecules produced by T suppressor cells (inducer and effector) mediate their effects in the presence of antigen without any requirement for the presence of specific MHC molecules (6,7). Tsi bear surface structures detected by anti-L3T4

*By "traditional" we do not imply "universally accepted", but only that it is a view held by many researchers in the field.

and anti-Ly-1 antibodies (like helper T cells), but unlike helper T cells, Tsi appear to react with anti-I-J antibodies (8). Thus, Tsi and helper T cells are distinguished by two criteria, a) surface phenotype, and b) mode of antigen recognition (requirement for specific MHC molecules).

In this discussion, we will concentrate on the Tsi and the antigen specific molecule mediating its function. The "traditional" view espoused above is illustrated in Figure 1. Note that the Tsi factor is composed of (at least) two chains, which may or may not be disulfide linked (6,7). Only one of these chains is responsible for the antigen specificity of the factor.

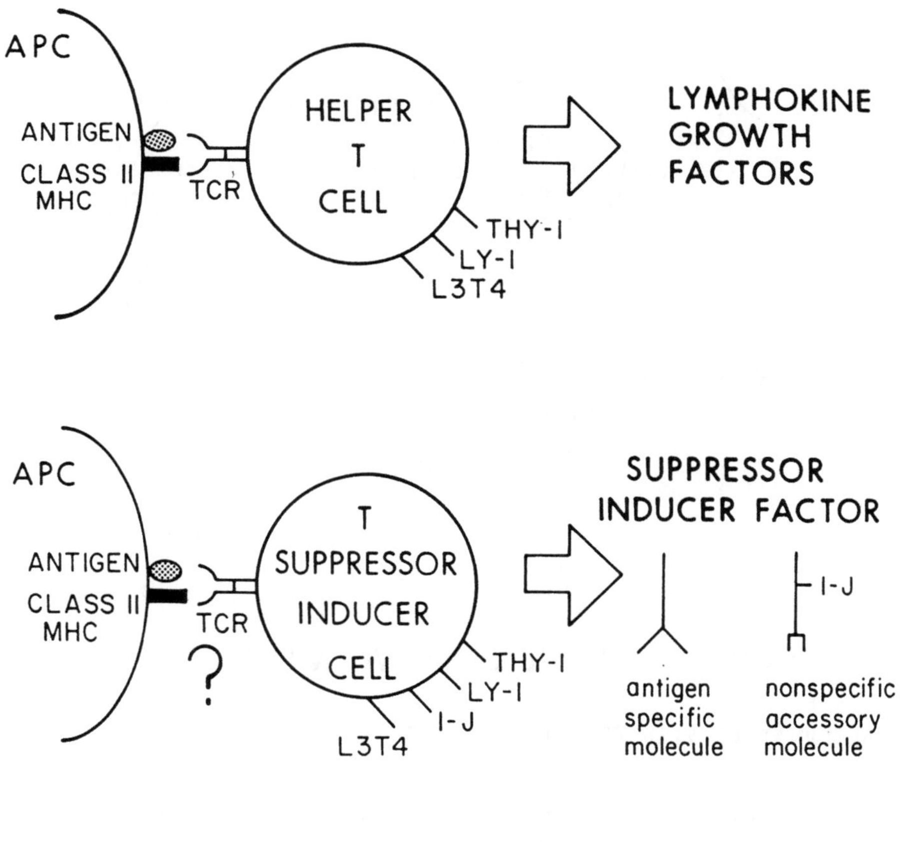

FIGURE 1. A "TRADITIONAL" VIEW

Recently, we have considered an alternative to this traditional view. Flood (9) observed that the antigen specific and the I-J$^+$ chains of the Tsi factor are made by different cells; that which produces the antigen specific chain bears a cell surface molecule profile indistinguishable from that of helper T cells. We therefore proposed the view shown in Figure 2.

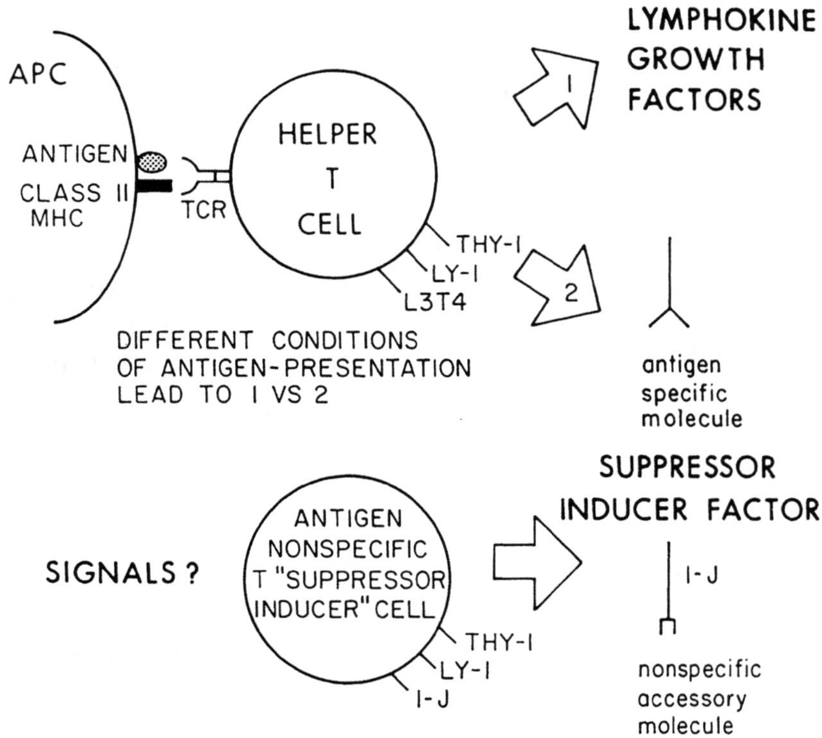

FIGURE 2. AN ALTERNATIVE VIEW

AN ALTERNATIVE VIEW OF HELP AND SUPPRESSION (Figure 2)

In this alternative view, the antigen specific molecule of the Tsi factor is actually made by (at least some) T helper cells. It is the interaction between the antigen presenting cell and the helper T cell which initially determines whether the cell will produce lymphokines (help) or antigen specific inducer molecules (suppression). This view presents an apparent paradox, however, revolving around the use of the term "recognize". That is, a single T cell (the helper cell) recognizes antigen in two distinctly different ways: a) in association with specific MHC molecules via the cell surface T cell receptor and b) in the absence of specific MHC molecules via the cell-free Tsi molecule. To appreciate this paradox, it should be noted that the T cell receptor heterodimer is necessary and sufficient for antigen plus MHC recognition (10) and that this Tsi molecule is apparently smaller (approx. Mr 25 Kd) than either chain of the heterodimer (11). In addition, functional Tsi have been described which lack one chain (β) of the cell surface T cell receptor (12).

Before suggesting a resolution of this apparent paradox (and discussing its ramifications), we must consider the validity of the model leading to the paradox. Evidence from two systems supports this alternative view of help and the induction of suppression.

HELPER T CELLS WITH Tsi FUNCTION

We have studied two well-characterized helper T cells and identified antigen-specific Tsi activity released from them (see Table 1). One of these cells is a helper T cell hybridoma (P18-68.8), that is, the product the fusion of a helper T cell and a thymoma cell (BW5147). The resulting clonal product of a single cell responds to I-A^d antigen-presenting cells (APC) plus the appropriate antigen (poly 18) by releasing lymphokines. The antigen, poly-18, is a nonrandom synthetic polypeptide of defined structure (13). Supernatants of the T cell hybridoma were found to contain a poly-18 specific Tsi activity with the characteristics shown in Table 1 (manuscript in preparation). By employing a series of closely related peptides and peptide variants of the poly 18 molecule we found that the Tsi molecule released from the cell had exactly the same antigenic fine specificity as displayed by the T cell in the T helper assay. The fundamental difference between the mode of response to the antigen by the Tsi molecule and the helper T cell was the requirement for I-A^d APC in the T helper system and the absence of such a requirement for the Tsi function (ie, the Tsi molecule worked equally well in assays utilizing H-2^b spleen cells).

Similar results were found for a T cell clone, D10.G4 (14). This cell line is maintained by lymphokines and antigen plus syngeneic, irradiated "feeder" cells. We found that supernatants of the clone contain an antigen-specific Tsi activity. In addition, we found that release of the Tsi activity required antigen presentation under conditions which do not trigger helper T cell function (15,16). For example, antigen presenting cells which have been irradiated with ultraviolet light (UV-APC) are poor stimulators of helper T cell function (17) but cause the release of antigen specific suppressor molecules (18). UV-APC, plus antigen, stimulate D10 T cells to release antigen specific Tsi molecules.

As yet, we do not know the relationship between the well-defined cell surface T cell receptor on helper T cells and the less well understood Tsi molecule. Moorhead, and colleagues (19) have shown that both bear similar serologic determinants. Further, De Santis, et al. (20) have implicated T cell receptor genes in the generation of suppressor T cell molecules. Nevertheless, more work is certainly needed to fully elucidate this relationship.

Based on the experimental evidence, we believe that the mechanism by which the Tsi functions antigen-specifically without specific MHC molecules and that by which T helper cells function antigen specifically only in the presence of specific MHC molecules, is, nevertheless, the same. The antigen fine-specificity experiments for the two functions of P18-68.8 argue for the identity of antigen-recognition structures on the T cell receptor and the Tsi molecule. How then, can this structure recognize antigen in such strikingly different ways?

RECOGNITION AND RESPONSE TO ANTIGEN BY T CELLS

Clearly, one problem requiring resolution is the use of the term "recognize". When we suggest that a helper T cell "recognizes" antigen plus MHC molecules, we mean that it binds to neither alone, and only responds to both (eg, by proliferation, lymphokine secretion, etc). The Tsi molecule, on the other hand, often can be shown to bind antigen (eg, by affinity chromatography) and responds regardless of the presence of specific MHC molecules (ie, it induces suppression).

TABLE 1

CHARACTERISTIC[1]	Tsi Function[2]		T helper Function[3]	
	D10.G4	P18-68.8	D10.G4	P18-68.8
a) Antigen specificity	Conalbumin	Poly-18	Conalbumin	Poly-18
b) MHC specificity	None	None	$I\text{-}A^k$	$I\text{-}A^d$
c) Requirement for $Ly\text{-}2^+$ T cells	Yes	Yes	No	No
d) Requirement for $I\text{-}J^+$ Chain	Yes	Yes	No	No
e) 14-30 Reactive	Yes	Yes	No(?)	No(?)

[1]Characteristics:
a) Antigen specificity in two assay systems.
b) Requirement for APC (or other cells) bearing appropriate MHC in each assay system.
c) Removal of $Ly\text{-}2^+$ T cells from assay system removes effect (characteristic of Tsi factors).
d) Effect seen only in presence of antigen-nonspecfic $I\text{-}J^+$ molecule from Tsi factor (characteristic of Tsi factors).
e) Cell or molecule bound by 14-30 antibody (specific for antigen specific molecule of Tsi factor (25)).

[2]Tsi function assessed as described elsewhere. Basically, sheep red blood cells (SRBC) coupled with relevant (or other) antigen are cultured with naive spleen cells in presence of Tsi factors. After 5 days, Tsi function is assessed as resulting inhibition of anti-SRBC plaque forming cell response.

[3]D10.G4 (14) and P18-68.8 (13) release lymphokines in presence of appropriate APC plus antigen.

These distinctions, however, are not absolute. Helper T cells have been observed to sometimes bind antigen (21) and in one case this has been demonstrated to be a direct interaction between the antigen and the T cell receptor (22). These antigen-binding helper T cells are only activated by antigen plus MHC. Similarly, some helper T cells, while responsive to antigen plus MHC, will respond to high concentrations of immobilized antigen alone (23,24). Thus, it is increasingly likely that the T cell receptor often is capable of "recognizing" antigen (without MHC) but that the presence of MHC allows the triggering of a measurable response.

Therefore, we suggest that the recognition of antigen by a helper T cell can lead, under one set of conditions, to production of lymphokines and, under a different set of conditions, to release of an antigen specific molecule whose specificity corresponds to that of the T cell. This molecule, unlike the cell itself, functions in the absence of specific MHC molecules. Helper T cells, under these latter conditions, favor the induction of suppression while at the same time fail to provide help. Thus, they contribute in redundant ways to the state of unresponsiveness to a specific antigen.

It is conceivable that this dual function is transient, that is, that a cell triggered to act as a Tsi for a period of time can be "frozen" into this state. This would happen if the genes encoding parts of the T cell receptor required for response (help) but not involved in antigen recognition were no longer expressed. The observation that Tsi hybridomas which constitutively produce antigen-specific molecules often lack effectively rearranged β-chain genes (12) might bear on this point.

HELP VS. SUPPRESSION AND OTHER IMMUNOLOGICAL DECISIONS

Thus, the helper T cell's interaction with the APC forms a decision-making process in the immune system, leading to either help or the induction of suppression. This interaction may well be under the influence of a number of extrinsic factors, some physical (such as UV irradiation), some hormonal, and some immunological. The APC, then, may serve as a data processor, but it is the antigen-specific T helper cell which is triggered into one or another regulatory pathway.

Multiple outcomes for a single interaction provides a source of semiosis in the immune system. Of course, the result of the interaction is "automatic", requiring nothing resembling human thought (or decision making). However, the sum of a system composed of such interactions, in which one signal in can equal different signals out might only be finally understood at a semiotic level. For, though we may understand the precise mechanisms of cellular interaction (as, say, for a single neuronal synapse) this does not, and can not, tell us how an idea is created, nor, for the immune system, how self and nonself are discriminated. Yet without an understanding of the precise mechanisms, any realization of higher level functions will be necessarily incomplete.

The decision point we describe may be central to the global decisions made by the immune system, though it is certainly not the only one. T cell control of delayed type hypersensitivity versus antibody responses, T cell control of immunoglobulin class switching, and the multiple regulatory effects of antibody and antigen-antibody complexes are all points at which a number of intrinsic and extrinsic factors seem to affect the outcome of the decision. The wiring of the immune response is complex, and what it is wired to do is only slowly becoming clear.

TOWARDS AN IMMUNOSEMIOTICS

We have implied that the presence of decision-making processes and one-signal-in : several-signals-out necessitates a semiotic system. This, however, is not how a semiotic system is defined. In order to address the semiotics of the immune system and to define a role of the type of decision process we envision, it would be useful to have a test for semiosis which does not depend upon soft definitions of creativity or intelligent behavior.

We suggest a test of semiotic systems proposed by Eco which depends on the principle of triadicity. A system in which a triad exists, even in the absence of human use (eg, the manipulation of the investigator) which cannot be reduced to a simple cause-effect diad, is semiotic (by this definition, of course). An example of such a nonreducible triad is the relationship between an object and its representation <u>and</u> the interpretation which maps one onto the other (in either direction, depending upon whether a message is being transmitted or received).

In order to understand the nonreducible triadicity of immune processes, it is first necessary to recognize an obvious, but important point: the immune system does not react to foreign invaders, per se, but rather to samples of these intruders which have already been effectively removed. For example, if bacteria infect a break in the skin, the presence of the bacteria does not, in itself, trigger a specific antibody response. Instead, non-specific processes (such as phagocytosis by macrophages) begin removing some of the bacteria in the course of normal wound healing. These ingested bacteria, or fragments of them similarly ingested, now may serve as signs of the infection, but only if they reside in an APC. Thus, molecules from the antigen processed by the APC now appear on the surface, together with class II MHC molecules. When the APC triggers specific helper T cells (if conditions are right) the helper cells, in turn, trigger specific B cells to make antibody. The antibody, finally, assists in clearing the bacteria (the original object). The immune system, then, has "interpreted" the presence of bacteria molecules on an APC to mean "bacterial infection" and has responded. Clearly, however, we might get the same interpretation if we administered only bacterial fragments, thus fooling the interpreter (and making vaccination possible). The fact that it is captured and not free antigen which triggers the immune response is so obvious that it tends to be overlooked as irrelevant. Yet it is this basic fact which converts the object to a sign and imparts the beginning of semiosis to the system (even at this simple level).

This is not to say, however, that it is the APC-T cell interaction which comprises a semiotic system. If we consider only the helper T cell function, then the interaction of an APC with a T cell is strictly a diad: cause-effect (ie, help). But if the immune system were a simple cause-effect diad (a defensive knee-jerk response) then there could be no such thing as self-nonself discrimination with specific responses to nonself.

When processed antigen on an APC acts as a sign of the presence of that substance in the body, this does not, in itself, dictate that a particular immune response will occur. Instead, there are a number of possible "interpretations" of that sign which may result in a number of different immune responses or in the suppression of an immune response (noting that such suppression may also be considered a response).

The type of decision-making process described herein, that of an APC-T cell interaction resulting in either help or suppression is a step towards understanding how the immune system interprets such signs. Of course, antigen on an APC is only one of many signs confronting the immune system, but these add to the richness of immune expression without altering its semiotic status.

ACKNOWLEDGEMENTS

The work described herein was supported by an Establishment Grant from the Alberta Heritage Foundation for Medical Research. D.R.G. is an Alberta Heritage Scholar.

REFERENCES

1. Schwartz, R.H. 1985. Ann. Rev. Immunol., 3:237.
2. Jensen, P.E. and J.A. Kapp. 1985. J. Mol. Cell. Immunol., 2:133.
3. Cairns, L., F.S. Rosen and Y. Borel. 1986. Eur. J. Immunol., 16:1277.
4. Haskins, K., J. Kappler and P. Marrack. 1984. Ann. Rev. Immunol., 2:51.
5. Dialynas, D.P., D.B. Wilde, P. Marrack, A. Pierres, K.A. Wall, W. Harran, G. Otten, M.R. Loken, M. Pierres, J. Kappler and F.W. Fitch. 1983. Immunol. Rev., 74:29.
6. Green, D.R., P.M. Flood and R.K. Gershon. 1983. Ann. Rev. Immunol., 1:439.
7. Dorf, M.E. and B. Benacerraf. 1984. Ann. Rev. Immunol., 2:127.
8. Eardley, D.D., D.B. Murphy, J.D. Kemp, F.W. Shen, H. Cantor and R.K. Gershon. 1980. Immunogenetics, 11:549.
9. Flood, P.M. 1985. Fed. Proc., 44:5216 (Abstr.)
10. Saito, T., A. Weiss, M.A. Norcross and R.N. Germain. 1986. VI Internatl. Congr. Immunol. Abstracts, p. 115 (Abstr. 2.21.7).
11. Beaman, K.D. and R.E. Cone. 1984. Biochem. Biophys. Res. Comm., 125:475.
12. Hedrick, S.M., R.N. Germain, M.J. Bevan, M. Dorf, I. Engel, P. Fink, N. Gascoigne, E. Heber-Katz, J. Kapp, Y. Kaufmann, J. Kaye, F. Melchers, C. Pierce, R.H. Schwartz, C. Sorensen, M. Taniguchi and M.M. Davis. 1985. Proc. Natl. Acad. Sci. (USA), 82:531.
13. Fotedar, A., M. Boyer, W. Smart, J. Widtman, E. Fraga and B. Singh. 1985. J. Immunol., 135:3028.
14. Kaye, J., S. Porcelli, J. Tite, B. Jones and C.A. Janeway. 1983. J. Exp. Med., 158:836.
15. Green, D.R., B. Chue, T.A. Ferguson, K.D. Beaman and P.M. Flood. 1985. Brit. J. Rheumatol., 24(Suppl. 1):105.

16. Green, D.R., B. Chue, T.A. Ferguson, K.D. Beaman and P.M. Flood. J. Mol. Cell. Immunol., (In press).
17. Pretell, J.O. In: J.A. Parish (ed), The Effect of Ultraviolet Radiation on the Immune System, Johnson and Johnson, 1983, p. 337.
18. Drebin, J.A., S. Schatten, A. Tominaga, S. Lefort, N.L. Letvin, R. Bast, S.B. Mizel and M.I. Greene. Ibid., p. 123.
19. Moorhead, J.W., R.L. Fairchild and R.T. Kubo. 1986. VI Internatl. Congr. Immunol. Abstracts, p. 337 (Abstr. 3.36.10).
20. De Santis, R., G. Palmieri, G. Doria and L. Adorini. 1986. VI Internatl. Congr. Immunol. Abstracts, p. 115 (Abstr. 2.21.8).
21. Rao, A., W. Ko, S. Faas and H. Cantor. 1984. Cell, 36:879.
22. Siliciano, R.F., T.J. Hemesath, J.C. Pratt, R.Z. Dintzis, H.M. Dintzis, O. Acuto, H.S. Shin and E.L. Reinherz. 1986. Cell, 47:161.
23. Walden, P., Z.A. Nagy and J. Klein. 1986. J. Mol. Cell. Immunol., 2:191.
24. Thomas, D.W. and M.J. Sovay. 1986. J. Immunol., 137:3401.
25. Ferguson, T.A. and G.M. Iverson. 1986. J. Immunol., 136:2896.

On I-J, a Network Centre Pole and AIDS

Geoffrey W. Hoffmann

Departments of Physics and Microbiology
University of British Columbia
Vancouver, B.C.
Canada, V6T 2A6

ABSTRACT

Basic features of the symmetrical network theory are reviewed briefly. Postulated differences in network connectivity between helper and suppressor cells are then related to the properties of the cell surface antigens they each express. A resolution of the I-J paradox is suggested in the context of the symmetrical network theory. A "centre pole" model for I-J is described, in which I-J is anti-anti-self. The theory includes the experimentally testable hypothesis that the difference in I-J between the two key strains B10.A(3R) and B10.A(5R) is not genetic, but rather reflects differences between self-stabilizing sets of T cell idiotypes, that are transmitted from generation to generation via a maternal influence on developing T cell repertoires.

Ziegler and Stites (1986) have published an autoimmunity theory of acquired immunodeficiency syndrome (AIDS). The I-J centre pole model leads to a network theory of AIDS, that is similar to the Ziegler-Stites theory, but a little more complete. It is suggested that the disease involves anti-anti-self and anti-anti-anti-self immune responses, resulting in a fundamental destabilization of the network. The theory provides several new experimentally testable predictions, including ideas on how to prevent the disease.

1. Introduction

The network hypothesis (Jerne, 1974) gave immunologists a new way of viewing the immune system. The symmetrical network theory (Hoffmann, 1980, 1982; Gunther and Hoffmann, 1982) is an attempt to relate many aspects of immune system behaviour to a small set of explicit postulates concerning the

NATO ASI Series, Vol. H23
The Semiotics of Cellular Communication in the Immune System
Edited by E. E. Sercarz et al.

system components. In this paper I will briefly review only some aspects of the theory. [The Contemporary Topics in Immunobiology network theory paper (Hoffmann, 1980) provides a more complete review.] I will then broaden the scope of the theory by discussing several phenomena that have not been discussed previously in this context: Lyt-2 and CD8 on suppressor T cells and on cells that recognize class I MHC antigens, a mechanism involving L3T4 and CD4 that results in low network connectivity of helper T cells, and the I-J paradox. The discussion of I-J leads into a network autoimmunity theory of AIDS.

2. Review of Basic Symmetric Network Theory

The symmetric network theory is based firstly on the fact that if an idiotype P is anti-idiotypic to an idiotype Q, then Q is also anti-idiotypic to idiotype P. We call this concept "first symmetry". According to this idea, both stimulation and killing should occur in both directions between P and Q. Eichmann and Rajewsky (1975) and Trenkner and Riblet (1975) first showed that idiotypes could be stimulated by anti-idiotypes, which was the "backwards" direction. Symmetry in stimulation can be understood in terms of processes that occur at the cell surface. It is a well established fact that the specific stimulation signal for both T and B lymphocytes involves the cross-linking of receptors. For instance, if a divalent idiotype P on one B cell can cross-link an idiotype Q on another B cell, then the converse is also the case. We have similarly been able to verify that killing is bidirectional (Cooper-Willis and Hoffmann, 1983). This was done using a model system involving two mutually specific hybridoma antibodies.

Of course not all V-V interactions are completely symmetrical. The interactions between an idiotype P on a B cell and a corresponding anti-idiotype Q on a T cell are obviously not symmetrical. Similarly, symmetry can be broken if P binds to Q plus an MHC molecule on one cell, while Q binds only to P on the other cell. Symmetry is nevertheless a key aspect of the network, and we seem to advance best if we first try to understand the network in terms of symmetric interactions, and subsequently try to understand the asymmetries, rather than beginning with the asymmetries.

Related to the idea of symmetry is the idea of multispecificity. Each V region recognizes and is recognized by multiple other V regions within the system. That is, each V region can interact functionally with a variety of other V regions that can be complementary to various parts of itself. There is no unique, privileged antigen binding site or paratope on the V region. [As a consequence of this idea, no additional postulates had to be made to account for H-2 restriction phenomena in this theory (Hoffmann, 1980), in contrast to some other theories of regulation (Cohn, 1981, von Boehmer et al., 1978).]

In our initial model building, we considered a drastically simplified system consisting of just four groups of clones: the B and T clones that recognize a particular antigen ("+" clones) and B and T clones that recognize those clones ("-" clones). For this system, a simple set of postulated interactions leads to a set of four stable states. We postulate mutual stimulation between positive and negative populations (based on the cross-linking of receptors mechanism), inhibition (blocking of receptors) by monovalent specific T cell factors, and reciprocal killing by antibodies and/or by specific killer T cells.

The four stable steady states of the system for a particular antigen are the virgin state (low levels of both plus and minus clones), the immune state (elevated plus, depleted minus levels), the suppressed state (elevated plus and minus populations) and the anti-immune state, which is the converse of the immune state (elevated minus, depleted plus). A simple mathematical model of the interactions shows that the four steady states can exist and be stable (Gunther and Hoffmann, 1982).

3. The Addition of IL-1 to the Model

The immune system is complicated, so we initially formulate models that are simpler in some respects than the real system, and which nevertheless have as many of the known system properties as possible. We know for instance that several non-specific factors seem to play roles in B cell proliferation and differentiation. We nevertheless begin our attempts to understand the properties of the system without taking them all into account; we anticipate that an understanding of many of the main phenomena will not involve all of these factors.

The first formulation of the symmetric theory included only specific signals to cells (Hoffmann, 1975). A non-specific factor that caused B cell differentiation to antibody secretion was later added to the model (Hoffmann, 1978), and increased its scope. We will see in this paper that including also the non-specific factor IL-1 further increases the scope of the theory. We will assume that T cell proliferation requires both a cross-linking signal and a non-specific factor (I-L1) from macrophages (A cells), while secretion of specific T cell factors results from just one signal, namely the cross-linking of receptors. [For B cells we assume the opposite, namely just one signal (cross-linking of receptors) for the stimulation of proliferation, and two signals (cross-linking of receptors plus a non-specific differentiation factor) for secretion of antibodies.]

4. Class I MHC Antigens, CD8 and Lyt-2, High Connectivity and Suppression

In our model the immune state is a state of low network connectivity, the virgin state is a state of intermediate network connectivity, and the suppressed state is a state of high network connectivity (Hoffmann, 1980). T cells that are predisposed to be suppressor T cells could be T cells whose relation to self antigens indirectly results in them having high network connectivity. (By high network connectivity I mean high connectivity to other V regions; see the note with the reference to my 1980 paper.) Conversely, T cells whose relations to self antigens result in low network connectivity would be predisposed to be helpers. An example of how this could come about can be seen by considering Lyt-2 bearing T cells in the mouse, and CD8 bearing T cells in man. These cells are involved both in the recognition of class I MHC and in suppression. Cells that express Lyt-2 have a better chance than other cells to interact functionally with class I MHC antigens, simply because the Lyt-2 antigen itself interacts directly with the class I antigens (MacDonald et al., 1982). Class I MHC antigens are the most ubiquitous cell surface self antigens, so T cells that recognize class I antigens could be stimulated to secrete specific T cell factors more than any other T cells. A cells are assumed in the theory to have receptors for specific T cell factors, and would therefore normally have more anti-class I MHC specific factors on their surfaces than factors of any other specificity. This would cause the preferential selection of anti-anti-class I MHC T cells, since T cells receive a proliferation signal (IL-1) from A cells. The selection of anti-anti-class I T cells means that the anti-class I, Lyt-2 bearing T cells would automatically have high network connectivity, as required in the theory for suppressor T cells. It is then not surprising to find that suppressor T cells typically express Lyt-2. The same considerations apply to CD8 T cells in man.

5. Class II MHC Antigens on A cells, CD4 and L3T4, Low Connectivity and Help

Helper cells express the L3T4 cell surface marker in the mouse and the CD4 marker in man. These marker have affinity for the class II antigens expressed on A cells (Biddison et al., 1985; Owens and Fazekas de St. Groth, 1987). T cells expressing these markers can therefore divide (receiving IL-1 from the A cell) relatively independently of any affinity for other V regions within the network. These T cells are then automatically low connectivity T cells, as required by the theory, in which low connectivity T cells are helper cells. Cells with V regions that recognize the class II MHC antigens on the A cell surface are preferentially selected, and the helper cells thus become class II MHC restricted. (In the next section I will suggest that other T cell V regions are also involved in the selection of these V regions, but to a lesser extent than for I-J bearing suppressor cells.)

6. I-J

The I-J phenomenon remains a fascinating puzzle (Tada et al., 1986, Waltenbaugh, 1986, Murphy, 1987). I-J is an important serologically defined marker (or markers) in mice associated mainly with suppressor cells, for which the gene cannot be found at the place where it was expected to be on the basis of genetic mapping using recombinant strains of mice. The two most important strains for defining I-J are B10.A(3R) and B10.A(5R). Definitive sequencing data now shows that there is no genetic difference between these two strains in the MHC chromosomal region to which the I-J difference between them maps (Kobori et al., 1986). Experiments with chimeras (Sumida et al., 1985, Uracz et al., 1985) and transgenic mice (Flood et al., 1986) have provided strong evidence for the idea that I-J determinants are V regions on T cells, that have been selected by class II MHC antigens (Schrader, 1979). The data do not provide direct evidence concerning the question of whether the I-J V regions are anti-class II MHC or anti-anti-class II, but we can consider the question of which possibility fits best in the context of the symmetrical network theory. I have argued above that anti-class II MHC cells should have low connectivity, corresponding to helper function, so it is difficult to see how I-J bearing T cells could also be selected on the basis of being anti-class II, and be able to act as suppressors. I will therefore consider how I-J bearing cells could be anti-anti-class II, have high network connectivity, and function as suppressors.

We might imagine two alternative topologies for the part of the network that is influenced by self class II MHC antigens, as shown in Fig. 1 and 2. The divergent topology of Fig. 1 is the conventional picture; we have heterogeneity in anti-class II clones (helper T cells), and even greater heterogeneity in the anti-anti-class II (suppressor?) clones. This model is not consistent with the idea that suppressor T cells have higher network connectivity than helper T cells, and there is no obvious basis for a common shape (I-J) being expressed on the V regions of all the anti-anti-class II cells. (A single monoclonal anti-I-J antibody is able to interact with V regions on many different suppressor T cells.) The alternative "network focussing" topology of Fig. 2 might seem unlikely at first glance, but it does fit together nicely with the ideas of the symmetrical network theory. This part of the T cell repertoire could be formed through a selective process involving both anti-class II (helper) cells and anti-anti-class II cells. An additional constraint for cells to be selected as helper cells, (besides having anti-class II specificity) would be that they should have a low affinity for as many anti-anti-self cells as possible, and similarly the constraint for selection of anti-anti-class II cells would be that they should recognize as many different anti-class II cells as possible. If such a selection process leads to the type of topology shown in Fig. 2, with more anti-class II cells than anti-anti-class II cells, the anti-anti-class II cells would have a higher network connectivity than the anti-class II cells,

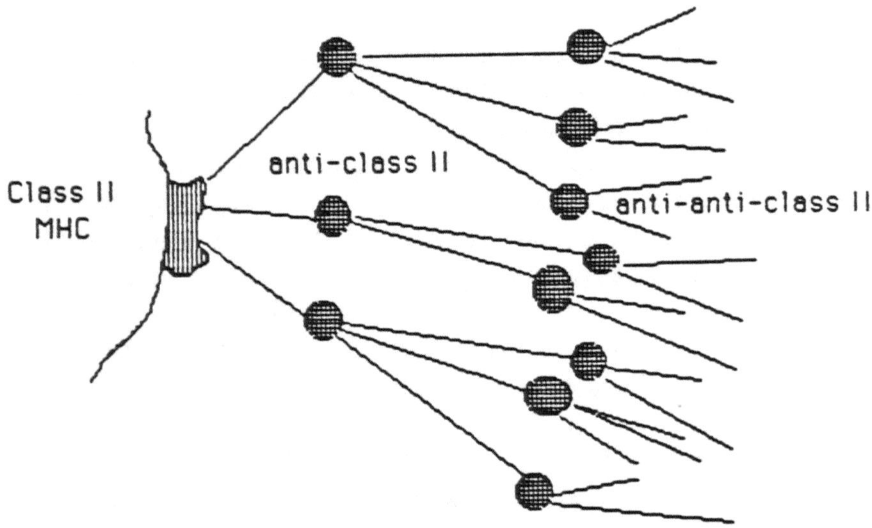

Figure 1. A divergent network topology for clones that relate to class II MHC self antigens.

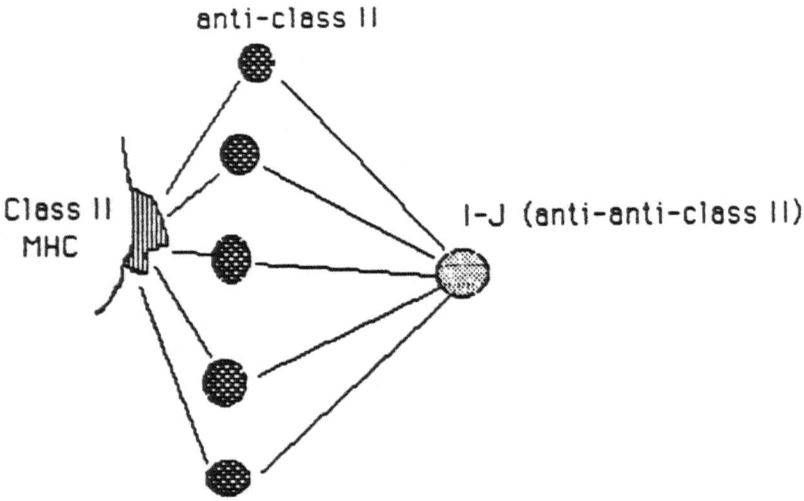

Figure 2. A network focussing topology for clones that relate to class II MHC self antigens.

so the anti-anti-class II cells could be suppressor cells and the anti-class II cells could be helpers. There could thus be a convergent selection process leading to the sharing of some shape by the anti-anti-class II cells, and that shape would be what we call I-J.

The convergence shown in Fig. 2 is not meant to imply that all the I-J bearing V regions are identical; I-J bearing specific T cell factors seem in many cases to have I-J determinants on one chain, and the other chain is free to have a specificity that relates directly or indirectly (idiotypically or anti-idiotypically) to the antigen.

The idea is that I-J bearing clones are selected on the basis of having an affinity for as many other V regions as possible. This could presumably be achieved by having a low affinity with each individual V region. If I seem to be stretching the notion of multispecificity to the limits of credibility, recall that the selection process involves both anti-class II and anti-anti-class II clones, so there is a lot of flexibility concerning which clones are selected. Clones that do not relate to other clones in the way shown in Fig. 2 would be at a selective disadvantage, and hence would not be selected.

A metaphor for the network is a circus tent, with I-J being the centre pole, and with mutual stabilization occurring between the centre pole and the rest of the structure. Each of the many other shapes interacts weakly with the central shape(s); other shapes are stabilized by their interaction with the I-J clones, and the I-J clones are in turn stabilized by the presence of the particular clones that have been co-selected. The centre pole holds up the canvas of the tent, and the canvas holds the centre pole in place. Both Fig. 2 and the centre pole metaphor are of course simplifications; I-J exhibits some heterogeneity, and as would be expected for idiotypes, not all anti-I-J reagents react with all I-J determinants (Waltenbaugh and Lei, 1984, Flood et al., 1986).

So far we have considered only V region interactions between the anti-class II and the anti-anti-class II T cells. This might be too simple. The V regions on, say, the anti-anti-class II T cells could also interact with other molecules on the helper cells, such as L3T4. But now both the V regions and the L3T4 molecules on the helper cells have complementarity to class II MHC (Owen and Fazekas de St. Groth, 1987). Thus both sets of molecules could help to cause the selection of I-J clones (fig. 2) that are anti-anti-class II.

7. A Hypothesis about the Difference between B10.A(3R) and B10.A(5R)

The centre pole metaphor emphasizes the interdependence of the shapes that are present, and the mutual stabilization idea leads to an unorthodox concept for the difference between the B10.A(5R) and B10.A(3R) strains, which is an important part of the I-J puzzle that needs to be resolved. Is it possible that there is no genetic difference between the 5R and 3R strains, or at least none that affects I-J? Could instead a selected set of mutually stabilizing shapes, including I-J, be inherited from generation to generation, with a strong maternal effect on the developing T cell repertoire? Perhaps the *I–J*^{b} phenotype can be stabilized in the presence of less than a complete set of *b* haplotype class II antigens, and perhaps the presence of less than a complete set of the *k* class II antigens is enough to stabilize the *I–J*^{k} phenotype once it has been established. If a different set of class II MHC antigens is present in an off-spring than in a mother, a particular constellation of maternal I-J V regions and anti-class II V regions could be unstable, but the presence of just one of the original class II genes might be sufficient to stabilize the presence of a particular set of anti-class II and I-J shapes.

There is ample precedent for the idea of a strong influence of a maternal idiotypic repertoire on the repertoire of off-spring. In the allotype suppression phenomenon a female is made immune to the allotype of a male, and their off-spring are then suppressed for the expression of the paternal allotype (Jacobsen et al., 1972; Herzenberg et al., 1973). There have also been reports that the immune state of the male can influence the immune state of the off-spring (Gorczynski and Steele, 1980; Cooper-Willis et al., 1985), and there is evidence that this effect takes place via material other than sperm in the ejaculate (possibly lymphocytes), that interacts with the mother's immune system (Gorczynski et al., 1983).

Eichmann (1975) has presented evidence that perturbations of an immune system by small numbers of suppressor cells can be passed from "generation to generation" in cell transfer experiments, and result in recurring modifications in the idiotypic repertoire of recipient mice. He injected as few as 10^5 spleen cells from mice suppressed for a particular idiotype into lightly irradiated syngeneic recipients, which then, after a long delay (6 weeks), also became suppressed for that idiotype. The newly suppressed mice could then be used as donors of spleen cells that could suppress subsequent "generations" of recipient mice for that idiotype.

The most striking evidence of profound maternal effects on T cell idiotypic repertoires was reported recently by Martinez et al. (1986). Immunization of adult female mice with an idiotype resulted in the lack of expression of that idiotype in the off-spring. Most significantly, this lack of expression was manifest also in the next generation. Similarly, in inbred strains of mice, the I-J phenotype of a female might have a profound effect not only on the I-J phenotype of her immediate off-spring, but also on the off-spring of her off-spring, and so on.

We still need to be able to explain how the difference between the 3R and 5R strains originated in the first place. Both strains resulted from recombinant events in crosses between B10 and B10.A mice. We would need to postulate that the original 3R mouse was the offspring of a B10 mother and a B10.A father, while the first 5R mouse was born of a B10.A mother and a B10 father.

The idea of a decisive maternal influence on I-J in 3R and 5R mice can be tested experimentally. Fertilized eggs from 3R mice ($I\text{–}J^b$) implanted into $H\text{–}2^k$ mice, should, by this hypothesis, give rise to $I\text{–}J^k$ off-spring. Furthermore, the offspring of the offspring should also type $I\text{–}J^k$. If they do, we will have a case of a long-term non-genetic influence on the apparent genotype, and a major part of the I-J paradox will be solved.

8. A Network Theory of AIDS: a Variant of the Ziegler-Stites Theory

It is commonly believed that the pathogenesis of acquired immunodeficiency syndrome (AIDS) involves an attack on CD4 (T4) cells by the AIDS virus, human immunodeficiency virus (HIV). In an important recent paper Ziegler and Stites (1986) have suggested that the HIV virus mimics class II MHC antigens, and causes the disease by inducing autoimmunity to self class II, together with an anti-idiotypic response against CD4 bearing cells. In the following I will suggest a model that is similar to that of Ziegler and Stites. It involves network interactions to a greater degree, and leads to additional testable predictions. The model includes immune responses to HIV, to the human equivalent of I-J, and to anti-host receptors on foreign cells. (The human equivalent of I-J has not been characterized experimentally, but I will assume it exists.) Fig. 3 shows the sequence of relevant cells, from cells that express self class II MHC to anti-anti-anti-anti-self.

We consider two complementary immune responses. The first is the response to HIV. The virus binds to the CD4 molecule on helper T cells. CD4 is believed to be complementary to class II MHC (Biddison et al. (1985); see Owens and Fazekas de St. Groth (1987) for analogous findings in the mouse). If CD4 is anti-self, then HIV is anti-anti-self, and anti-HIV is anti-anti-anti-self. Now anti-anti-anti-self is directed against the equivalent of I-J. Hence these antibodies can potentially destabilize the network by attacking the "centre-pole".

The second response is an anti-anti-self response, that occurs when lymphocytes are transferred from one person to another. We recently showed that the injection of fully allogeneic lymphocytes into mice (say strain A cells into a strain B mouse) routinely causes an anti-anti-self (B anti-(A anti-B)) humoral response in addition to the conventional B anti-A response (Hoffmann, Cooper-Willis and Chow, 1986). The anti-anti-self response is against the receptors of the foreign lymphocytes that recognize the host. Homosexuals can receive allogeneic

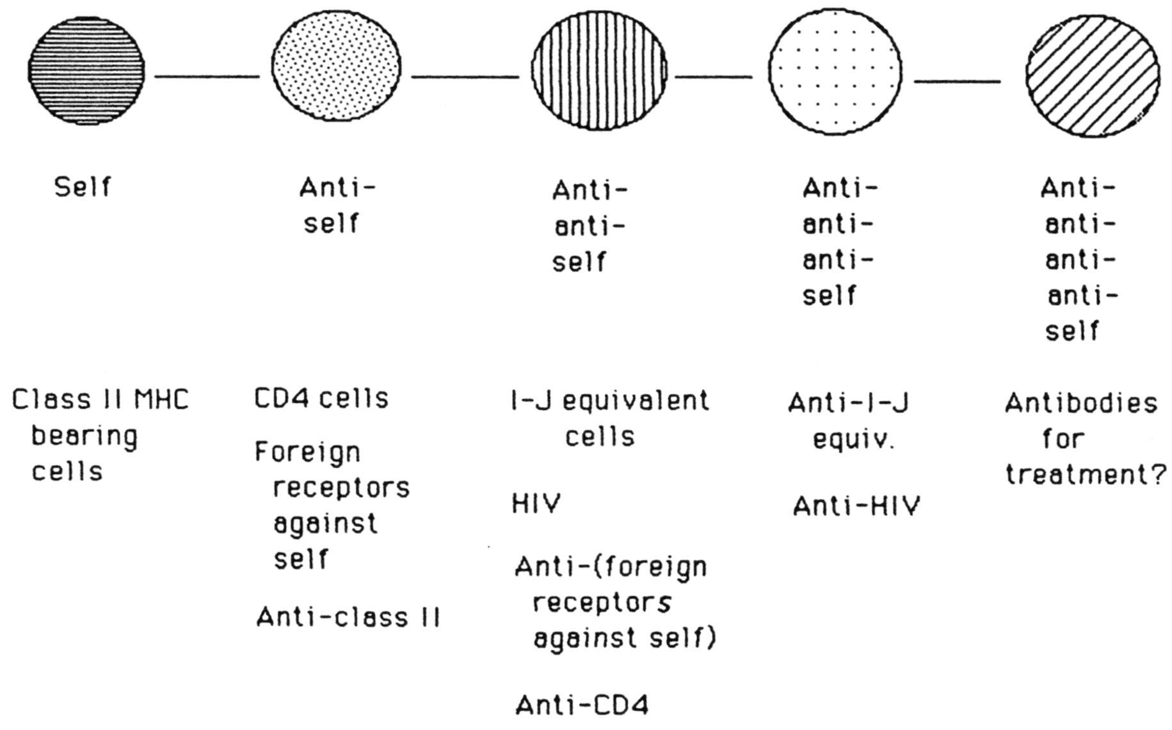

Figure 3. The sequence of cells and specificities from self class II MHC to anti-anti-HIV.

lymphocytes together with sperm in ejaculates, and this could result in anti-anti-self responses. Other high risk groups (intravenous drug users, recipients of blood transfusions) also typically receive cells when infected.

One can imagine that the anti-anti-self and anti-anti-anti-self responses could synergize with each other disastrously. We know that an effective way of making anti-idiotypic antibodies is to inject immune complexes. It seems that particularly strong immune responses are evoked when we confront an immune system simultaneously with a shape and an approximately complementary shape. A catastrophic destabilization of the network by the above pair of responses, which would include an attack both on the centre pole and on CD4 cells, seems quite possible. One result might be the emergence of of a competing "I-J". A competition between two different would-be centre-poles could incapacitate the entire network.

It is interesting that antibodies specific for CD4 lymphocytes, that could be anti-anti-self class II, have been reported to be present in the serum of AIDS patients (Dorsett et al., 1985; Kiprov et al., 1984; note however a conflicting report by Williams et al., 1984).

These ideas suggest several experiments, some but not all of which were suggested by the Ziegler-Stites theory. Firstly, we clearly need to develop sensitive assays for detecting anti-anti-self and anti-anti-anti-self (anti-centre pole) antibodies in AIDS patients' sera. It will incidentally be interesting to look for such antibodies also in patients with other immunological disorders. On the basis of our "centre pole" ideas, we recently looked for and found anti-anti-self antibodies in auto-immune MRL strain mice (Kion and Hoffmann, in preparation).

We do not have antibodies to I-J in man, so we need to do some of the experiments in the mouse system. If L3T4 (the equivalent of CD4) is anti-class II MHC in the mouse, and I-J is anti-anti-class II, then we might find that anti-L3T4 (anti-anti-class II) would bind to anti-I-J (anti-anti-anti-class II). Similarly, anti-HIV is assumed to be anti-anti-CD4, so it could bind to anti-CD4. The theory predicts that anti-HIV antibodies in people that develop AIDS will react with the small population of I-J equivalent suppressor T cells.

Immunization with a combination of anti-L3T4 and allogeneic cells is predicted to induce AIDS in mice. The anti-L3T4 antibodies would play the role of HIV, inducing an anti-anti-anti-self response, and the allogeneic cells would induce an anti-anti-self response.

The theory suggests that rather than inducing immunity to HIV, we should be trying to find ways of making people tolerant to it. Alternatively, we need to devise forms of immunization that evoke responses that do not endanger the centre pole.

If the anti-anti-self response to allogeneic cells is a key component in the development of the disease as postulated, then it is possible that most anti-HIV responses that occur in the absence of allogeneic cells will be harmless.

Idiotypic characterization of anti-HIV antibodies might be useful for prognosis; not all anti-HIV responses are expected to be both anti-HIV and anti-centre-pole, but all the ones that are anti-centre pole are predicted to cause the disease. By analysing the idiotypes of the anti-HIV antibodies, we might be able to tell whether someone who is seropositive is at high risk of contracting the disease. We might furthermore expect that the anti-HIV idiotypic profile is quite narrow in diseased patients, since the selection process that eventually gives rise to the disease is slow and presumably stringent.

For people who are sero-positive and are at high risk for contracting the disease, an appropriate treatment could be with anti-idiotypic antibodies against the anti-HIV antibodies. We see in Fig. 3 that these are anti-anti-anti-anti-self antibodies. The same treatment could be applicable for patients that already have the disease, but it might be very difficult or impossible to restabilize a completely destabilized network.

Acknowledgement

This work is supported by the Medical Research Council of Canada, Grant No. MA-6770, and the Natural Sciences and Engineering Research Council of Canada, Grant No. A-6729.

Notes and References

Biddison, W. E., Rao, P. E., Talle, M. A., Goldstein, G. and Shaw, S. (1984). Possible involvement of the T4 molecule in T cell recognition of class II HLA antigens. Evidence from studies of CTL-target cell binding. J. Exp. Med. 159, 783-797.

Cohn, M. (1981). Pinch hitting for Hans Wigzell. In "Immunoglobulin Idiotypes and their Expression", ICN-UCLA Symposia on Molecular Biology, vol. XX, Janeway, C., Sercarz, E. E., Wigzell, H. and Fox, C. F., editors, Academic Press, New York, pp. 831-849.

Cooper-Willis, A. and Hoffmann, G. W. (1983). Symmetry of Effector Function in the Immune System Network. Mol. Immunol. 20, 865-870.

Cooper-Willis, C. A., Olson, J. C., Brewer, M. E. and Leslie, G. A. (1985). Influence of paternal immunity on idiotype expression in off-spring.

Immunogenetics, 21, 1.

Dorsett, B., Cronin, W., Chuma, V. and Ioachim, H. L. (1985). Anti-lymphocyte Antibodies in Patients with the Acquired Immune Deficiency Syndrome. Amer. J. Med. 78, 621-626.

Eichmann, K. (1975). Idiotype Suppression. 2.Amplification of a Suppressor T Cell with Anti-Idiotypic Activity. Eur. J. Immunol., 5, 511-517.

Eichmann, K. and Rajewsky, K. (1975). Induction of T and B cell immunity by anti-idiotypic antibody. Eur. J. Immunol. 5, 661-666.

Flood, P. M., Benoist, C., Mathis, D., Murphy, D. B. (1986). Altered I-J phenotype in E_α transgenic mice. Proc. Nat. Acad. Sci. USA, 83, 8308-8312.

Flood, P. M., Waltenbaugh, C. W., Tada, T, Chue, B., Murphy, D. B. (1986). Serologic heterogeneity in I-J determinants associated with functionally distinct T cell regulatory factors. J. Immunol., 137, 2237-2244.

Gorczynski, R. M., Kennedy, M., Macrae, S. and Ciampi, A. (1983). A possible maternal effect in the abnormal hyporesponsiveness to specific alloantigens in the offspring born to neonatally tolerant fathers. J. Immunol., 131, 1131.

Gorczynski, R. M. and Steele, E. J. (1980). Inheritance of acquired immunological tolerance to foreign histocompatability antigens in mice. Proc. Natl. Acad. Sci. USA, 77, 2871.

Gunther, N. & Hoffmann, G. W. (1982). Qualitative Dynamics of a Network Model of Regulation of the Immune System: A Rationale for the IgM to IgG Switch. J. Theoret. Biol. 94, 815-855.

Herzenberg, L. A., Chan, E. L., Ravitch, M. M., Riblet, R. J., and Herzenberg, L. A. (1973). Active suppression of immunoglobulin synthesis. III. Identification of T cells as responsible for suppression by cells from spleen, thymus, lymph node and bone marrow. J. Exp. Med., 137, 1311.

Hoffmann, G. W. (1975). A theory of regulation and self-nonself discrimination in an immune network. European J. Immunol. 5, 638-647.

Hoffmann, G. W. (1978). Incorporation of a non-specific T cell dependent factor into a network theory of the regulation of the immune response. In "Theoretical Immunology", G. I. Bell, A. S. Perelson and G. H. Pimbley (Eds.) Marcel Dekker, New York, pp. 571-602

Hoffmann, G. W. (1980). On network theory and H-2 restriction. In Contemp. Topics in Immunobiol. 11, 185-226. Note that due to an undetected printer's error (half a sentence was omitted) this paper contains the erroneous statement (p. 207) "T cells [with] low connectance would preferentially suppress", which should read "T cells with low connectance would preferentially help, and those with high connectance would preferentially suppress." I now use the term connectivity instead of connectance in this context, because connectance has been used by mathematical ecologists in a different sense, and it is sensible to use two different words for two different meanings (Hoffmann 1981, 1982). If we have a network consisting of N different clones, and the strength of interaction between clones i and j is K_{ij}, then the connectance is the fraction of the N^2 parameters K_{ij} that are non-zero, while the connectivity of a particular clone i is

$\sum_{j=1}^{N} K_{ij}x_j$, where x_j is the population size of clone j.

Hoffmann, G. W. (1981). On network theory, Ly phenotypes and connectivity. In "The Immune System. Festschrift in Honor of Niels Kaj Jerne on the Occasion of his 70th Birthday." C. M. Steinberg and I. Lefkovits, editors, S. Karger, Basel, vol. 1, pp. 28-34.

Hoffmann, G. W. (1982). The application of stability criteria in evaluating network regulation models, in "Regulation of Immune Response Dynamics", C. DeLisi and J. Hiernaux, eds., CRC Press, Boca Raton, Florida, pp. 137-162.

Hoffmann, G. W., Cooper-Willis, A. and Chow, M. (1986). A new symmetry: A anti-B is anti-(B anti-A), and reverse enhancement. J. Immunol. 137, 61-68.

Jacobson, E. B., Herzenberg, L. A., Riblet, R. J. and Herzenberg, L. A. (1972). Active suppression of immunoglobulin allotype synthesis. I. Chronic suppression after perinatal exposure to maternal antibody to paternal allotype in (SJL x Balb/c) F1 mice. J. Exp. Med. 135, 1163.

Jerne, N. K. (1974). Towards a network theory of the immune system. Ann. Immunol. Inst. Pasteur 125C, 373-389.

Kibrov, D. D., Busch, D. F., Simpson, D. M., Morand, P. R., Tardelli, G. P., Gullett, J. H., Lippert, R., and Mielke, H. (1984). Antilymphocyte serum factors in patients with acquired immunodeficiency syndrome. in "Acquired Immune Deficiency Syndrome", M. Gottlieb and J Groopman, Eds., published by Alan R. Liss, New York.

Kobori, J. A., Strauss, E., Minard, K. and Hood, L. (1986). Molecular Analysis of the Hotspot of Recombination in the Murine Major Histocompatibility Complex. Science 234, 173-179.

MacDonald, H. R., Glasebrook, A. L., Bron, C., Kelso, A. and Cerottini, J.-C. (1982). Clonal Heterogeneity in the Functional Requirement for Lyt-2/3 Molecules on Cytolytic T Lymphocytes (CTL): Possible Implications for the Affinity of CTL Antigen Receptors. Immunol. Rev. 68, 89-115.

Martinez-A., C., Toribio, M. L., De La Hera, A., Cazenave, P. A. and Coutinho, A. (1986) Maternal transmission of idiotypic network interactions selecting available T cell repertoires. Eur. J. Immunol. 16, 1445-1447.

Murphy, D. B. (1987). The I-J Puzzle. Ann. Rev. Immunol., in press.

Owens, T. and Fazekas de St. Groth, B. (1987). Participation of L3T4 in T cell activation in the absence of class II major histocompatability complex antigens. Inhibition by anti-L3T4 antibodies is a function both of epitope density and mode of presentation of anti-receptor antibody. J. Immunol. 138, 2402-2409.

Schrader, J. W. (1979). Nature of the T cell receptor. Scand. J. Immunol. 10, 387.

Sumida, T, Sado, T., Kojima, M., Ono, K., Kamisaku, H. and Taniguchi, M. (1985). I-J as an idiotype of the recognition component of antigen-specific suppressor T-cell factor. Nature, 316, 738-741.

Tada, T., Uracz, W. and Abe, R. (1986). Are there Unique I Region Controlled Determinants on T Cells? In "Paradoxes in Immunology" (G. W. Hoffmann, J. G. Levy and G. T. Nepom, Eds.), CRC Press, Boca Raton, Florida, 1986, pp. 253-265.

Trenkner, E. and Riblet, R. (1975). Induction of Antiphosphorylcholine Antibody Formation by Anti-idiotypic Antibodies. J. Exp. Med. 142, 1121-1132.

Uracz, W., Asano, Y., Abe, R. and Tada, T. (1985). I-J Epitopes are Adaptively Acquired by T cells Differentiated in the Chimaeric Condition. Nature 316, 741.

von Boehmer, H., Haas, W., and Jerne, N. K. (1978). Major histocompatability complex linked immune responsiveness is acquired by lymphocytes of low responder mice differentiating in the thymus of high-responder mice. Proc. Nat. Acad. Sci. U.S.A. 75, 2439-2442.

Waltenbaugh, C. (1986). I-J - Immunoregulatory Molecules in Search of a Gene. In "Paradoxes in Immunology", (G. W. Hoffmann, J. G. Levy and G. T. Nepom, Editors), CRC Press, Boca Raton, Florida, 1986, pp. 271-281.

Waltenbaugh, C. and Lei, H. Y. (1984). Regulation of immune responses by I-J gene products as detected by anti-I-J monoclonal antibodies. J. Immunol. 133, 1730.

Williams, R. C., Masur, H. and Spira, T. J. (1984). Lymphocyte-Reactive Antibodies in Acquired Immune Deficiency Syndrome. J. Clinical Immunol. 4, 118-123.

Ziegler, J. L. and Stites, D. P. (1986). Hypothesis: AIDS is an autoimmune disease directed at the immune system and triggered by a lymphotropic retrovirus. Clin. Immunol. and Immunopath. 41, 305-313.

IMMUNE SUPPRESSION: AN INTEGRAL PROPERTY OF THE IMMUNE SYSTEM?

NA Mitchison
Imperial Cancer Research Fund Tumour Immunology Unit,
Department of Zoology and Cell Biology
University College London, Gower Street, London WC1E 6BT, UK

Who sits at the T table, and how are they getting on with their careers? Well it's always gratifying to meet the helper cell and his cousin the cytotoxic cell, both now at their prime, both with an excellent reputation for their benevolent work (in defending against disease), and both wearing as shiny buttons their newly discovered receptors. No problems there. And then there's the antigen presenting cell also doing very well. He's a Billy Bunter-ish figure, very greedy and has his buttons smeared with peptides. He too has his own accepted place at the table, and there is no hint of scandal in the offing. But there's another member of the family with a far less satisfactory character. She's been around for quite a time and doesn't seem to be getting on nearly as well as had been hoped. In a Victorian sort of way she's deemed to have past her prime. In molecular circles she's considered disreputable, and the Yuppies won't talk to her at all. What makes it worse is that she's still having a lot of fun, and is not short of besotted admirers (me for instance). I refer, of course, to the suppressor T cell

Most of this essay on suppressor T cells concerns the immune suppression genes (abbreviated to Is genes), that control the activity of suppressor T cells and that provide the only firm foundation for the whole edifice of suppression. We discuss three questions about these genes: why are they, what are they, and where do they act. We shall conclude that they and the cells that they control are functionally important and interesting in their evolution, but that they are not fundamentally distinct from other genes and cells. We shall be left with the null hypothesis that suppression reflects the uniqueness of the context within which they function rather than the uniqueness of the cells themselves.

This meeting has an anti-molecular mood, and so we are on

NATO ASI Series, Vol. H23
The Semiotics of Cellular Communication in the Immune System
Edited by E. E. Sercarz et al.

the look out for properties of the immune system that resist interpretation in terms of yet another set of molecules. The null hypothesis leaves us with suppression as a convincing candidate for such an integrative property.

Why are Is genes?

My text for this question is a leper with swollen nerves who will soon die. Most such people belong to the borderline-tuberculoid part of the leprosy disease spectrum, and in terms of evolutionary fitness they have had rotten luck. Just a little more cellular immunity and they might have thrown off their infection; a little less, and they might have been protected against tissue damage, as R de Vries and T Sasazuki (throughout this essay only senior investigators are mentioned, although they had many colleagues) have shown, by Is genes. Down where the Is genes hang out, the host and its parasite enjoy quite a cosy relationship: there is little tissue damage and a better chance of survival for the host on the one hand, and on the other multiplication of the bacilli that gives them a good chance of spreading to another host. This constitutes a feed-back loop selecting for Is genes, in which the parasite sets up antigens that generate immunopathology, and this in turn selects for Is genes that benefit the parasite. Generalizing, chronic infection will tend to select Is genes, just as acute infection selects contrary immune response (Ir) genes that favour the immune response.

This view contradicts the prevailing "immunological orchestra" view of suppression which sees the immune response as controlled by a balance between plus and minus forces, simply because that is in the nature of homeostatic control. That is a bland and boring view that places all antigens on the same footing. No doubt chronic infection is not the only force selecting for Is genes; allergy and auto-immunity may play a part as well, although probably only a minor one. The point is that evolutionary selection must be important, not only from what we expect from selection but also from what we know of naturally occurring Is gene polymorphisms.

Thus to the evolutionist the Ir and Is genes constitute a coded image of the world in which the species has lived during its evolution. An all-wise comparative anatomist could read much of the history of the vertebrates from inspection of, say, the brain: in the corpora quadrigemina for instance she would see our amphibian ancestors with their gigantic optic tecta, staring from the mesozoic slime. Just so would an all-wise immunologist learn of past plagues by inspecting present Ir and Is genes. In fact this is not analogy but identity: what else are those who propose to sequence the entire human genome but the ultimate comparative anatomists? More prosaically D Oliveira and I predict that past exposure to, say, malaria will be found to perturb significantly DQ gene frequencies (Parasitology Today, in press).

What are Is genes?

The answer to this question is both highly informative and somewhat paradoxical: the major Is genes are simply class II genes of the major histocompatibility complex. The paradox lies in the fact that these class II genes also function as Ir genes so that evidently the same gene-product can control both helper and suppressor T cells. To some extent this paradox is relieved by two supplementary pieces of information: (i) each individual has several of these class II genes, some of which specialise in suppression, and (ii) within the class II-controlled subset of lymphocytes that used to be regarded as the helper subset, some cells evidently specialise in helping B cells while others specialise in helping other T cells (including suppressor T cells), as is further discussed below.

What is so informative is that class I genes of the major histocompatibility complex do not show up as Is genes. That is surprising, as the suppressor T cells which directly mediate suppression ("suppressor effector cells") bear the T8 marker which is a hallmark of class I gene control. So what is going on? It's very unlikely that class I gene control occurs without showing up in the genetics, for these genes are known to be at least as polymorphic as class II genes. The best explanation is that these cells are under idiotypic (or "network") control, a circumstance under which class I gene intervention is no longer mandatory. To be precise, experimental studies indicate that T cells can be driven by idiotypic interactions either with or without the intervention of class I gene products, depending presumably on the extent to which that interaction fills the T-cell receptor's binding site. At any rate the idiotypic control hypothesis is already supported by observations that Tada and Taniguchi (independently) have made on suppressor cells in allogeneic chimeras.

How good is the evidence for suppression-specialised class II genes? Ideally, in man the genetic analysis should proceed in the following three stages: first a "disease association" links a particular disease to a particular gene polymorphism; next, in family studies a single dose of the gene confers susceptibility to infectious disease (or resistance if the disease depends on hypersensitivity), "sib concordance"; finally monoclonal antibody to the class II gene product proves able to enhance the immune response in vitro, "monoclonal antibody relief". In the mouse the analysis should proceed along similar lines, although there the experimenter can look directly at immune responses without proceeding via disease susceptibility; in addition recombinant inbred lines substitute for families and make research easier. In practice of course in neither man nor mouse is the evidence often so complete, but at least it is increasing.

This summer at the International Congress of Immunology D Oliveira and I summarised the evidence for Is gene effects congregating selectively at the E locus in the mouse and at DQ in man. Since then the list of instances has increased, from 6 to 8 in the mouse (through the addition of another set of minor transplantation antigens by N Nanda, and of Leishmania minor antigens by E Liew) and from 4 to 5 in man (through the addition of leprosy by T Sasazuki). So far there have been no exceptions to the rule that, given a choice between A and E, mice always use the latter to control suppression. Sometimes there is no choice: about a fifth of mice do not have properly functional E genes, and then they either switch suppression to A, the other class II gene (eg perhaps in the response of C57 mice to lysozyme) or else loose suppression (eg in the response to Trichinella). Neither E nor DQ function exclusively as Is genes, as both also serve as Ir genes in certain responses. The fact that this specialisation of function is not exclusive, coupled with the fact that E and DQ are not evolutionarily homologous (their DNA sequences suggest that it is DQ and A which share a common origin) clearly show that Is function is to an extent labile.

The list of 5 instances of Ir function in man has an interesting composition. Three of them involve chronic infection (tuberculosis, leprosy and schistosomiasis), one involves intermittent infection associated with immunopathology (streptcocal) and one involves allergy (cedar pollenosis). This is the sort of distribution that would be expected if chronic infection is indeed the principle force selecting for Is genes. Of course the sample is by no means random, and it is far too early to draw a firm conclusion. It is also significant that some antigens, such as F protein in our own studies, seem to attract the activity of Is genes, while others that have been quite well studied such as cytochrome C do not do so. This confirms the idea that Is genes are selected for particular purposes, rather than serving a general "orchestral" role.

Where do Is genes act?

Evidence has been growing that T4-marked, class II controlled lymphocytes divide into two groups that help respectively B cells and other T cells. Antibodies such as Leu-8 in man, Ox-22 in the rat, and Qal in the mouse make that kind of division. These monoclonals and allo-antisera have been picked up in various empirical screens, and at the time I reviewed the subject at the last International Congress they had little rational basis. This has changed dramatically, thanks largely to studies of the molecular biology of the T200 molecule carried out by J Trowbridge and N Barclay.

They discovered that alternative splicing of transcripts of the T200 gene generates markers that distinguish these two groups of lymphocytes from one another. The proteins encoded by the gene are expressed at the cell surface in the form of four peptides of size 220, 205, 190, and 180 KD. The splicing events affect the N-terminal domain of the protein that lies distal to the cell membrane. Many monoclonal antibodies have been made against human T200, and over 32 of them recognise the unspliced domain of the protein proximal to the membrane. These antibodies are classified as belonging to the CD45 group, bind to all four T200 peptides, and mark all the T4 cells. In addition another group of over 15 monoclonals recognise only the 220 and 205 peptides, (and sometimes also the 190 one) and they are classifed as CD45R. Finally one monoclonal (named UCL1) recognises only the 180 peptide, and that one I propose should be classified as CD45R'.

Meanwhile the cell biology has been sorted out, by D Mason for CD45R in the rat, by S Schlossman for CD45R in man, and by P Beverley for CD45R' in man. They discovered that four T cell populations are involved: suppressor-effectors (Tse), cytotoxic cells (Tc), suppressor/cytotoxic-inducers (Tsci), and B cell-helpers (Th), with phenotypes, Tse:T8, CD45 unknown; Tc:T8, CD45R or R'; Tsci:T4, CD45R; Th:T4,CD45R' (Fig.1). Thus the T4 set splits into two with concordance between the CD45 markers and function. It is important to note that in this classification cytotoxic cells and suppressor-effector cells not only bear the same markers but are also controlled by the same type of helper cell.

This information does not tell us clearly whether Tsci and Th are separate lineages or merely represent distinct activation states of the same cell. Two considerations favour the latter view. One is direct: my colleague P Beverley has obtained evidence that C45R'-negative cells acquire this marker upon entering the memory pool. The other is less direct but perhaps stronger: alternative splicing smells strongly of modulation rather than irreversible cell committment. When cells differentiate irreversibly they normally do so by switching the transcription of genes on and off. When differentiation is reversible (ie modulation) they engage in manoeuvres which fall short of this, and involve mechanism such as altering the transcript from a given gene, as used by B-cells in respect of their heavy-chain immunoglobulin genes when they differentiate into plasma cells. The alternative splicing of T200 transcripts clearly falls into the latter category.

Are suppressor cells distinct from other T cells?

To summarise the argument so far:

(i) In terms of ability to generate suppression, not all antigens are equal. Among the small number of antigens that happen to have been chosen thus far for detailed experimental analysis this is the case. And this is just what consideration of selective pressures in evolution would predict.

(ii) Likewise not all parts of the major histocompatibility complex are equal. Certain class II molecules are particularly effective in mediating suppression, although they evidently do not concentrate on this task exclusively. Furthermore this function seems to transfer from one class II molecule to another during the course of evolution, and even within species as a result of loss mutation.

(iii) The T cells that respond to antigen-presentation by these molecules form a distinct group as judged by surface markers. However the distinction seems to be a transient property, reflecting a phase of activation rather than irreversible committment.

(iv) The T cells that mediate the final act of suppression by inactivating helper cells resemble cytotoxic T cells, in their surface markers and in the nature of the cells that activate them. This suggests that although these two groups of effector cell have evolved for different purposes they may use the same mechanism and belong to fundamentally the same cell type.

To these we may add a further consideration that was evident at the time of the International Congress, and that I shall not attempt here to justify further:

(v) Suppressor T cells are more highly connected than other lymphocytes T or B; that is, idiotypic binding of their receptors plays an extraordinarily important part in their interactions with other cells. The I-J markers and VH restrictions (both are aspects of what may be termed "epirestriction") are hallmarks of this high connectedness.

All these properties of the suppressor system could be explained in terms of one or more molecular mechanisms unique to the immune system. That is what most of the specialists working in the area hope, and let nobody doubt the power of the molecular fix in the history of immunology. Some of us may have believed that the T cell receptor would turn out to be just another immunoglobulin, but we were proved wrong. So today the suppressor receptor may turn out to include the newly discovered gamma chain, or even an entirely novel molecule.

I would argue that the weight of the evidence now supports the alternative assimilationist view. In terms of the five points listed above, it can be expressed in the following terms: if you take the right antigen and present it via the right MHC molecules to the right T cells, then it will activate through an idiotypic interaction a cytotoxic cell that in turn (again through an idiotypic interaction) inactivates a helper cell.

The major weakness of this view is that the hypothetical cytotoxic reaction has yet to be demonstrated. Its major strength is that many of what seemed to be unique features of suppression have either blown away or been revealed to occur elsewhere in the immune system: I-J, V_H-restriction and antigen-specific factors are all examples.

One final question: need the assimilationist assume that everything happens via selection of the secondary T cell repertoire? In other words, during the development of an individual immune system does a portion becomes focussed on idiotypes, and does this in turn select particular MHC molecules as presenting elements, determine the degree of activation and so on? This seems to me oversimplified. Surely in the evolution of the immune system something much more complex has taken place, in which MHC genes, V genes, and the selective processes involved in acquiring a secondary repertoire have all co-evolved under selective pressure from parasites and other sources. All of these systems have their own languages, and that is why a framework of semiotics is needed for their understanding.

SIGNS AND MEANING IN THE STUDY OF THE IMMUNE AND HIGHER NERVOUS SYSTEMS

P. A. Bretscher,
Department of Microbiology,
University of Saskatchewan,
Saskatoon, Saskatchewan,
Canada, S7N OWO.

"There is a delicate form of the empirical which identifies itself so intimately with its object that it thereby becomes theory."

Goethe, quoted by J. Berger in About Looking.

Introduction

Semiotics is the study of how people use signs to communicate. It seems likely that significant semiotic findings will be incorporated with time into a neurophysiological theory of how we perceive the world through our senses, make sense of it, and use signs to communicate our understanding. Signs can be received through any or several of the five senses, and their meaning is distinct from what they are. For example, a bunch of roses, the signifier, can express passion, the signified, if an interpretant is present who recognizes the intended relationship, the sign, between bunch of roses and passion. Interpretations are dependent on the context in which they are given; in other words, a bunch of roses does not always signify passion (1).

Sassure, who was active at the turn of this century, is recognized as a founding father of Structuralism, the attempt to understand what characteristics and properties of a language (in the broadest sense) are essential to its function of conveying ideas and information. His concepts arose as a reaction against the accepted views of his time, accoring to which language could be used to objectively describe objective findings. It was then generally accepted, for example, that each word exists by itself and has objective meaning. The roots and history of words were emphasized, i.e. their diachronic (from the Greek meaning along/across time) aspect, whereas Sassure emphasized in addition the

NATO ASI Series, Vol. H23
The Semiotics of Cellular Communication in the Immune System
Edited by E.E. Sercarz et al.

relationships between the words of the language, as exemplified by their use at a particular time, i.e. their synchronic aspect. From his viewpoint, language is in part a collection of words. The meaning of each, not only their definition, is dependent on the meaning of other words. The whole network of words represents a paradigm. Two concepts discussed by Sassure will be important for our purposes. Firstly, the very process of perception or observation is not objective but contains an inherent bias which affects to a significant degree what is perceived. This can be objectively demonstrated by analysing the sounds that are distinguishable by adult people speaking different languages. Objectively similar sounds can be more easily distinguished by an adult if a language he is familar with employs this "phenomic" distinction to discriminate between similar sounding words that have different meanings, but the two sounds are much less easily distinguished if not employed for such discrimination. An objective "phenomic" distinction between sounds, if "significant" in the language, is easily recognized as a "phonetic" distinction. This example illustrates the generalization that objective differences not endowed with "significance" in our language, or even the paradigm we employ to understand some aspect of the world, are less easily perceived than those endowed with such significance. Secondly, Sassure pointed out that a language must have formal structural features that are accepted by two people communicating; utterances will obey laws reflecting these structural features, and hence be interpretable. He likened the formal features to the rules of chess, and an utterance to a particular game. A formal rule of English determines that there is a difference in meaning between "boy kicks girl" and "girl kicks boy" (1).

Structuralist and semiotic concepts have been very influential in fields as diverse as literary criticism and anthropology. To my mind "structuralist" thinking in the context of immunology means at least a recognition that determining what happens in particular and well defined experimental situations is not enough for gaining an insight into the immune system; an effort has to be made to select those situations that are significant, and significance can only be assessed by judging whether the situation bears upon the general properties of the immune system. I am certainly in sympathy with this as far as it goes. Some would disagree and maintain that a thorough description of many particular situations, especially if carried out at the molecular level, that has well established laws and procedures for investigation, will reveal the general properties of the immune system and the mechanisms responsible for these properties. This view is based on the judgement or premise that observations made in a particular setting can be uniquely interpreted without a consideration of findings in other settings, a view in complete disagreement with the essentials of structuralism.

The study of stimulus-response reactions, e.g. how antigen interacts with its cellular receptors and the chemical consequences of such an interaction, are essential to immunology. I shall discuss later the sense in which some of the most intriguing attributes of the immune system cannot be accounted for only in terms of such reactions. Ehrlich realized that the great specficity of the immune system, and its ability to respond to virtually all foreign antigens, means it must also have the potential to respond to self antigens, a potential that must not (in general) be realized. This requirement for self-nonself discrimination led to the compelling hypothesis that the continuous presence of self antigens (from some time early in ontogeny) is responsible for inhibiting the induction of anti-self reactivity. This example of the immune system's adaptability, an attribute that provides the means for essential physiological regulation, cannot be explained solely as the result of a single stimulus-response reaction, but at best by a few such reactions strung together in a meaningful way. It has become clear in recent years that cells interact in multiple and complex ways, and that the immune system is exceedingly plastic and adaptable. This complexity and adaptability must serve some physiological needs. How and why are cells sometimes suppressed and sometimes induced? How and why are such decisions dependent on the past history of the animal? Does an identification of certain physiological needs, such as self-nonself discrimination, provide us with a handle on how cell intereactions are organised to produce physiologically desirable regulation? What cellular structures or products record the immunological experience of the immune system? How do these molecules influence whether antigen interacts with its specific cells (belonging to a particular subclass) to induce, paralyse or suppress them, i.e. how are they used as meaningful signs? I think these most interesting questions are unlikely to be answered by studies limited to stimulus-response reactions of the immune system. Once fully described, it will be possible to split the immune system into a set of stimulus-response reactions, but this way of looking at things neglects the importance of the manner in which the stimulus-response reactions are interconnected to give rise to such physiological attributes as self-nonself discrimination. An approach more likely to be successful in revealing the mechanisms responsible for the general attributes of the immune system is one whose aim is to understand the physiological action of important regulatory molecules and how their activity is controlled. From this perspective, the regulatory molecule can be regarded as the signifier, the conditions under which they are activated or produced, being physiologically crucial, as the signified, and the target cells, on which they act, as the interpretant. For example, the conditions under which some class of regulatory molecule is activated or produced could be just those that result in the induction of an aggressive response against the antigen; in this case, the regulatory molecules, the signifiers, express foreigness, the signified, and they act on target cells, the interpretant, thus acting positively (directly or indirectly) on the production of the aggressive immune response. Thus a set of stimulus-response reactions can be strung together in a manner that makes biological sense, and it is consequently appropriate to say the set has meaning.

The potential relevance of semiotics to understanding the general attributes of the immune system in terms of cellular interactions led to the organization of the workshop on Immunosemiotics. Some topics for discussion were suggested by the organizers, including the relationship between semiotics, the immune and higher nervous systems. The workshop stimulated my thinking not only about immunology but also the relationship between the theory of knowledge and semiotics and their potential relevance for defining some required functions that the nervous system must perform in the process of perception and theory making, or constructing paradigms of explanation. These thoughts clarified some of the problems I had faced as I attempted to grasp the possible relevance of semiotics to immunology.

Epistemology, semiotics and neurophysiology

A valid solution to the problem of self-nonself discrimination has been a major aim of almost all immunological theories. Similarly, a recognition of the attributes of higher nervous systems is not only important in itself but because attempts to account for these attributes are a major stimulus to neurophysiological theory. A biologist is likely to regard semiotics and epistemology, the study of the basis of knowledge, as important but transient fields. Significant semiotic and epistemological findings will in time be incorporated into a description of how neural cells interact with the environment and each other to give rise to the various attributes we recognize as thought and the ability to communicate.

Theory making is a form of communication with oneself, and one not capable of thought will not communicate with others. We first employ the majority of signs we use to communicate with others as tools in the formation of our own thoughts. The process of sign formation is thus closely related to the process by which we make sense of the world.

The British empiricists of the 17th and 18th centuries were the first philosophers to lay great stress on the idea that our knowledge of the real world is crucially dependent on observation (2). This viewpoint, so natural to us today, was a very significant advance and was inspired, no doubt, by the realization that the extensive observations of Kepler were necessary for Newton to develop his mechanics. (I think it could be argued that structuralism arose as a reaction against a view that payed too exclusive a homage to hard nosed reductionism and empiricism, a view influenced by Kepler's success.) There are two reasons why a consideration of Hume's philosophy (3), the most thoroughly extreme and purest expression of this empiricism, is relevant to what I wish to discuss here. The starting point for Hume is an analysis of the stuff of mental experience, or perceptions as he calls them, and how they might be used in the formation of memory and concepts. His analysis provides an appropriate context for discussing some aspects of memory and sign formation. Secondly, Hume's analysis does not fulfill his intuitions, and his assumptions concerning the basis of knowledge lead to a dead end. This impasse requires one either to accept that knowledge is impossible, a conclusion Hume came to with much reluctance and uneasiness, or a diagnosis of where his analysis might be modified and the impasse overcome.

My understanding is that Hume wished to provide a justification for empirical science and to discredit metaphysics. His philosophy arose as a reaction to a tradition in which metaphysics had for centuries been inevitably employed to arrive at a proof or description of God, the entity responsible for all things throughout time and space. God's characteristic attributes were his permanence and unchangeability, i.e. he was in some sense outside space and time. These metaphysical discourses contrasted with recent developments in science, developments that could not be ignored by an intelligent person of broad sympathies. I believe Hume wished to stress the empirical nature of knowledge and to thereby justify science whilst discrediting metaphysics. His analysis, however, suggested to him that knowledge was impossible. In this sense it was incredible and did violence to his intuitions.

Hume suggests that introspection results in the realization that there are two kinds of perception, the more vivid being impressions and a consequence of immediate perception (of entities in the external world), and less vivid ideas. These ideas represent memory and/or imagination. Hume appeals to his reader to examine his own perceptions and asks him whether he does not find his own ideas to be limited in the following sense. Ideas may be complex, but can be broken down into simple elements; elemental ideas are only possible if the corresponding elemental impression has occurred, though novel complex ideas, corresponding to a novel combination of previously experienced elemental impressions, can arise. Hume thought, I believe, that the door to metaphysics would be open once ideas not closely related to experience were accepted as valid.

It had been a common and implicit assumption of metaphysics that logic alone, perhaps with the aid of some observations for illustrative purposes, could give rise to a meaningful description of reality. Today it is recognized that logic is an essentially negative criterion. It allows one to rule out inconsistent and therefore impossible descriptions of reality, but it does not provide us with a positive criterion for what reality is. Hume initially felt that a description of reality, as it occurs in science, could be generated from logic and observation alone; yet, in the process of critically trying to justify this view, he realized that other principles, of which the principle of induction was one that he recognized, were required, and he could not justify this principle solely on the basis of logic and observation. In other words, a person endowed only with a logical capacity and a sensory apparatus would not try to make sense of the world at all. To do so he would need some principles, dis-tinct from logical ones, which would provide him with the inspiration or criteria to organise his perceptions such that he could develop a paradigm for interpreting the world. I shall refer to these principles as alogical as they are neither required nor disallowed on logical grounds.

Some of these alogical principles are very elemental and become apparent if we pay attention to where Hume's analysis does not ring true, namely in his analysis of imagination and particularly memory. An adult is put into a position A, a door is opened and he is shown a room in which a table stands. The next day he is shown the same room with the same table but from a different position, position B. He is asked whether the table seen today is the same as yesterday. The adult who correctly recognizes the table is asked why he believes it to be the same table. He may reply, for example, that both tables had rectangular tops, were about the same height, and were highly polished. These answers are likely even if the table top, when seen from positions A and B, form different irregular quadrilaterals, the positions are different distances from the table, and the table was seen on the two occasions under different light conditions. It usually happens that someone <u>remembers</u> the table to be rectangular far longer than he remembers the literal shape of the table top he has seen, i.e. the irregular quadrilateral. This is clearly different from what Hume envisaged. According to Hume ideas are pretty good copies, though less vivid, of the corresponding impressions. The example of the recognition of a table shows that an observer's impressions are often transformed in a particular way to be recorded as ideas that are quite different from the impressions from which they are derived. Three aspects of this transformation are interesting. Firstly, the memory (of a property of the object) is encoded as a <u>transformation</u> of a constellation of impressions; the resultant recorded idea attempts to provide a description of the object that is <u>independent</u> of the <u>viewer's position</u>. The example just given, in which an adult recognizes the table, illustrates that this transformation is in practice necessary if memory is to be useful. The recognition of tables is a typically academic exercise, but the importance of recognition is clearly crucial for survival; it allows animals to recognize, and attempt to avoid, a preying foe. How such transformations and consequent recognition are achieved neurologically is a fascinating question. A more simple and particular form of this question is well known. How can a shape be recognized when situated in a quite separate part of the visual field from when it was first seen? Recognition occurs even though quite distinct rods and cones are stimulated on the two occasions.

The ability to transform impressions into useful ideas must be a highly complex process, and is related to other basic attributes of the nervous system. For example, our observer saw a whole room, and yet he had developed a way of transforming his impressions into a series of discrete entities or objects. It would seem that the resolution of the mass of impressions into distinct objects, and the transformation of the impressions derived from one entity into useful ideas (of rectangularity and surface properties) must be mutually dependent. Suppose there was a book on the table when it was seen the first time, but not the second. The recognition of the table requires a recognition that the book was not an integral part of it. This example also illustrates the second important aspect of the transformation process, namely its intimate connection to the process of categorization. A person seeing two "quadrilateral" tables might realize them to be both rectangular and hence, in this respect, classify them together. Classification is not trivial in the least; it embodies what we are led to believe or think is significant.

A basic epistemological question closely related to the nature of transformation and classification is the relative contribution of the external world and the structure of our brain to these processes. Consider a reality that is a series of different pictures, each picture being made up of regularly spaced but differently coloured dots, the dots being placed to give the impression of a flower. Compare this reality to the one in which the same kind of regularly spaced coloured dots are randomly placed with regard to one another. In the former case, petal shape is likely to be recognized as a constellation of recurrent impressions and hence become an idea. This really means that the category of "petals" has formed. This categorization would have to entail the recognition that shape defines petalness, the orientation of the shape being unrestricted, i.e. having no significance, and its colour only slightly so (i.e. not green). The recognition of the categories of leaves, stems and so forth, would allow further laws to be discovered. Leaves but not petals are found very close to stems. No such categories and laws are possible from perceiving the pictures of randomly placed but regularly spaced coloured dots. The only laws that could be deduced would be at a different level, i.e. the world consists of five elementary dots, each a different colour, and these are randomly placed at regularly spaced grid points.

The picture we now have of memory is radically different from Hume's. Significant memory, as distinct from passive and objective recording, requires there to be some regularities in nature and hence the impressions we receive. Our brain contains pattern recognition devices that allow the recognition of these regularities. Recognition of petalness is due to pattern recognition in the sense I use it here; a shape in two dimensions is recognized independently of its orientation. A considerably more remarkable ability develops with experience, namely the ability to transform some of these regularities to objectively describe our experience. This process could be described in some cases, such as the identification of the table as rectangular, as pattern recognition in three dimensions. A major class of regularity allows us to form the impression of entity or object; a constellation of continuously changing but similar and related impressions "moves" together if the table is moved. It seems inevitable that the ability to form the idea of a table being rectangular, when by objective perception its top makes an irregular quadrilateral shape, must involve the hypothesis that the table is one entity and perceiving it from many different angles. A description then arises which is constant in time and independent of viewing position. This ability of the brain is not only necessary for significant memory and the formation of categories but must in the long run tell us much about the processing of visual information. The brain functions of pattern recognition and meaningful transformation of constellations of impressions into distinct ideas represent alogical principles that were not recognized by Hume. They appear to be necessary for the simplest kinds of thought.

Thirdly, these alogical principles have definite semiotic and structuralist facets. Consider two kinds of individual, person No. 1 who memorizes perfectly his impressions, i.e. acts like a camera and records the "quadrilateral" table top, and person No. 2, whose memory is an idea that is derived by transforming a constellation of impressions, i.e. he remembers a rectangular table top. Person No. 1 would not group different constellations of impressions in any significant way, and his memory would be objectively perfect. Person No. 2 would have the ability to put the table into various categories, such as wooden, rectangular and shiny object, or whatever categories he had developed as being significant. Person No. 2, on seeing a table that is not quite rectangular, is likely to record it as rectangular and quickly forget his original impression. This is surely an example of Sassure's realization that perception is not objective. Secondly, person No. 2, but not person No. 1, can attempt to communicate. He does not transmit his impressions (in the Humean sense) but his ideas. His ideas are not copies of his impressions but impressions transformed in a manner that makes sense to him. The nature of the transformations are representative of his paradigm, and so the ideas have meaning and are signs.

A realization of the semiotic significance of transformation came to me from a remark by Eco made during the meeting. The remark struck me forcefully, though I was unable immediately to understand why. He pointed out that signs were often visual, but rarely if ever olfactory. Smell does not have a well defined extension, and we remember it either literally as an impression, or we forget it. We do not transform it. We therefore cannot use smells as signs in the sense used above, i.e. transform the impression of a particular smell in a way to express significance. Our sense of smell, if leading to memory, is recorded as literally and objectively as possible, i.e. the olfactory sense is similar to the visual perception of person No. 1 above. The memory of smell retains the particularity of the impression. This retention makes smell particularly suitable as a sign for recalling the experience of particular places and particular occasions. Smell is a <u>private</u> <u>nostalgic</u> <u>sign</u> by virtue of its <u>association</u> with other impressions. Smells can be very evocative for some, though I understand this is not universally so.

This is not the place to attempt to list all the different alogical principles relied upon in theory making. There are a few further points I should like to make. Categorisation can occur at successive levels. The categories of wooden tables, chairs and spoons lead to the category of wooden objects with relatively complex but regular and defined shapes. This is a subcategory of man made objects. (I imagine Ohno's contribution to this book will discuss in part how categories can evolve.) Secondly, categories are useful both in ordering our experience of the world and in detecting anomalies or inadequacies in our viewpoint. It is of course well appreciated that the recognition of a paradox is often a turning point of real progress. For example, the existence of regular wooden objects in an area uninhabited by man could give rise to speculation of an archeological nature. Leonardo da Vinci, on finding a fossil seashell on a mountain, surmised the mountain had once been under the sea. Our sensitivity to paradox is in part responsible for the strange power produced by surrealistic art, in which inappropriate signs are put together, e.g., the fusion of inanimate and living material.

The impulse to describe our perceptions in terms of ideas that are "objects", i.e. entities that change only little with time, and whose "description" is independent of the position from which they are seen, is not only a requisite for categorisation and useful memory, but is hypothetical in nature; I hypothesise that a "rectangular" table is responsible for my perception of a table with an irregular quadrilateral top. The "rectangular" table is the reality behind my perception. Physics consists of laws that describe how fundamental entities interact. Conservation laws, of mass, energy and momentum, for example, govern these interactions. Fundamental entities and laws do not change with time, and a criterion of their fundamental nature is that they are valid everywhere. They are outside space and time and are the reality behind our perceptions. It seems to me that physics is a much more elaborate but similar phenomenon as the phenomenon of the rectangular table being behind the quadrilateral one; in both a reality outside space and time is held to be responsible for our perceptions. I think it no accident that metaphysicians found God, the ultimate for reason for everything, as existing outside space and time. Our brain struggles to find invariances.

Lastly, the resolution of our perceptions into entities suitable for categorization and memory formation are surely insufficient as a foundation for describing all mental behaviour. One regulatory function of the brain has not been touched upon though it has a very significant semiotic function and also bears on an awkward aspect of Hume's analysis. Hume distinguised impressions, the immediate result of perception, as more vivid than ideas, these constituting memory and imagination. It seems to me that vividness is too simplistic a distinction between these two kinds of perception. It is clearly most important that we do not confuse our thoughts with direct perception. Jung has described a situation where such judgement was lacking. A mountaineering patient had recurrent dreams of climbing a ridge and walking off an imaginary path into space. The patient continued to climb against Jung's advice, and was reported to have killed himself in a manner forcast by his dreams. Isn't self consciousness in part a regulatory signal telling us, to put it into Hume's terms, when our perceptions are ideas and not impressions, i.e. allowing us to distinguish between memory and imagination from direct perception? Our consciousness expresses our knowledge of whether our perceptions are derived from the outside or inside worlds. We may be impelled to run when we come across a lion, and in this case it is self consciousness that allows us to imagine a lion at leisure.

Immunosemiotics

I shall not attempt to define in what sense the immune system might be usefully described as semiotic, but rather illustrate by my discussion the kinds of questions a semiotic perspective should bring to the fore. We already know that complex signals occur between cells, and some of the structures involved in this signalling. An understanding of these signals can exist both at the descriptive level of what they are, and at the higher level of their role in, or contribution to, the organization of the cellular interactions that gives rise to the general attributes of the immune system. Our knowledge at these two levels will be fused once our understanding is complete, and we shall then know the significance of every detail of the signalling processes between all subclasses of lymphocyte for the functioning of the immune system as a whole. In the meantime, the relationship between details of the signalling process and the attributes of the immune system, or the sign value of the particulars of cell interactions, raises two kinds of question: what restrictions may the recognized attributes of the immune system impose upon the signalling processes between cells, and does what is believed to be true at the cellular/molecular level make sense in terms of the general attributes of the immune system? If not, we should either be suspicious of the validity of what we know, or accept that they reflect attributes of the immune system not yet recognized; in this case we should try to discover what such attributes might be. These two kinds of question can be regarded as asking what restrictions does the higher level organization (recognized as attributes) impose on the lower level of cellular interactions, and whether what is known at the lower level is commensurate with what is known at the higher level. I shall discuss some questions belonging to both categories. I begin by outlining some of the different kinds of organization that most probably occur at the higher level and potential means for distinguishing them. The ability to distinguish regulation that serves different physiological requirements is most important. Some of the major apparent inconsistencies between conclusions drawn from different experimental systems is probably because the experimental systems studied correspond to different physiological states. Distinct regulation must be responsible for the five attributes listed below.

(1) Self tolerance as the negative component of self-nonself discrimination

Nature: specific unresponsiveness to self antigens as a consequence of their presence before and after the immune system has attained competence. Characteristics: not associated with a state of immunity (i.e. activated effector function) of any kind; unresponsiveness affects the induction of all classes of immunity.

(2) Feedback regulation

Nature: specific negative regulation preventing an uncontrollably large response to an antigen; requires a monitoring system to assess when immunity is of sufficient strength to deal with the pathogen so that the inhibition of the induction of further immunity is not disadvantageous. Characteristics: negative regulation for a particular class of immunity; the negative regulation is induced and expressed under those conditions that lead to the induction of the particular class of immunity whose further induction it inhibits.

(3) Positive memory as seen in the secondary response

Nature: positive regulation, associated with a previous response, thus allowing a more rapid response on reinfection with the same, or an antigenically related, pathogen. Characteristics: these are more broad than would be anticipated on straightforward considerations, For example, memory for the humoral response can be expressed even in animals that have not previously made a substantial antibody response upon first exposure to the antigen. This indeterminateness does not lead to confusion as the secondary response is the only form of positive regulation so far on the list.

(4) Negative memory associated with the secondary response and the phenomenon known as original antigenic sin

Nature: the inhibition of the induction of a response to certain epitopes of an antigen as a consequence of priming with a related antigen; the epitopes not responded against are present on the challenging but not the priming antigen. Characteristics: successful inhibition of the response to these epitopes appears to be associated with the successful mounting of a secondary response to the other epitopes. The phenomenon is best characterised in antibody systems and first came to light in the response to variants of influenza virus responsible for successive epidemics (4).

(5) Exclusivity between classes of immunity

Nature: the induction of different classes of immunity is exclusive and the conditions under which they are induced are different. It was originally shown that the induction of antibody to an antigen specifically inhibited the induction of DTH, and this phenomenon was given the same of immune deviation (5). Later studies revealed the more general phenomenon in which the induction of cell mediated immunity, in the form of delayed-type hypersensitivity (DTH), is associated with unresponsiveness for the induction of antibody (6), or oral immunisation can produce IgA immunity and systemic unresponsiveness for a variety of other classes (7). I use the term immune deviation in the broader sense. Characteristics: specific and negative regulation for one class of immunity induced and expressed under conditions that result in the induction of an opposing class of immunity.

I examine below how the existence of these attributes may in some cases restrict the possible signs used in intercellular communication, and problems I believe to exist when we try to incorporate some molecular/cellular models into the higher level description of the immune system.

(1) The potential role of helper T cells as the sign system required for self-nonself discrimination

I assume, as most others would, that self-nonself discrimination is dynamically maintained even in the adult. This is supported by the occasional development throughout life of destructive autoimmunity. Since no intrinsic property distinguishes foreign from self antigens, a sign system must exist to provide this discrimination when both are present, i.e. when a foreign antigen impinges upon the immune system. In recent years immunologists have employed as antigens many proteins (e.g. insulins and cytochromes) derived from animals closely related phylogenetically to the immunized adult animal. Helper-like T cells specific for the foreign but not the shared, self epitopes are induced in this manner. These studies demonstrate that in the adult the specificity of these helper-like T cells define the universe of foreign epitopes by virtue of their specificity for them. Thus the binding of these molecules to an antigen can be used as a sign that the antigen is foreign and a response should be mounted against it. The (relative) absence of such helper-like T cells specific for self epitopes ensures self antigens cannot induce a response. The <u>specific receptors</u> or <u>factors</u> of these helper-like T cells, the <u>signifiers</u>, define <u>foreigness</u>, the <u>signified</u>; the <u>precursor cells</u>, the <u>interpretant</u>, have appropriate cell surface structures and internal biochemistry such that they are induced only when specific helper T cells (or specific factors derived from them) are present. This is essentially a restatement of the theory of self-nonself discrimination, proposed some years ago, according to which the induction of all precursor cells is helper T cell dependent (8). (The question of whether there are different kinds of helper T cells is not crucial to their potential role as signifiers of foreigness.) The pivotal role of helper T cells in this hypothesis highlights the importance of the conditions under which they are themselves induced. Consistent with the hypothesis is the recent demonstration that the primary induction of precursor helper T cells requires specific effector helper T cells that act via the linked recognition of antigen (9).

(2) Chronic immune responses and autoimmunity: Is there a positive side to original antigenic sin?

It is generally accepted that chronic stimulation of the

immune system by an antigen that induces antibody and crossreacts with a self antigen may induce antibody to the self antigen. Helper T cells that bind to foreign epitopes on the foreign antigen help the induction of newly generated B cells specific for both the foreign and self antigen (8).

The potential induction of autoimmunity by antigens cross-reactive with self raises a further question. Could another layer of regulation have developed to minimize the induction of autoimmunity while not significantly interfering with the response to the foreign antigen? There appear to be few T cells specific for self at any one time (10), but chronic stimulation by a crossreacting antigen has the potential for inducing newly generated precursor cells specific for the self antigen. It would be advantageous from this viewpoint if specific inhibitory elements were induced during a chronic response that acted upon virgin but not memory precursor cells; this would allow the chronic response to proceed but inhibit the induction of newly generated precursor cells, some of which would be specific for the self antigen (11). The existence of original antigenic sin, as seen in the humoral response to variants of influenza virus during successive epidemics, and in analogous experimental models (12), suggests such regulation occurs. Note that the specific inhibitory elements discussed here are quite different from the suppressor T cells proposed by some as being involved in maintaining self tolerance. Such suppressor T cells are usually thought of as being specific for either epitopes of normal self antigens, or for idiotopes of lymphocyte receptors that are themselves specific for other self epitopes (13). The physiological role of the inhibitory element discussed here would be to prevent the induction and spread of autoimmunity during chronic infections, but the recognition elements would be <u>specific</u> for <u>foreign epitopes</u>, and act preferentially on <u>virgin</u> as <u>opposed</u> to <u>memory</u> cells.

(3) What criterion is employed in determining the class of response induced by antigen?

I have described a theory that proposes both why it is advantageous for the immune system to differentially regulate the induction of cell-mediated and humoral responses in the manner observed, and how such regulation is achieved (14,15). I shall describe the theory here only as far as required to provide a context for some further semiotic considerations, and I make no attempt at experimental justification. The theory deals with the diffential regulation of cell-mediated immunity (CMI) in the form of delayed-type hypersensitivity (DTH), IgM and IgG antibody. I shall for reasons of simplicity only consider cellular antigens.

The unifying idea of the theory is that the existence of these three classes of immunity, and their differential regulation, together provide a system for effectively fighting foreign invaders and simultaneously minimizing the consequences of any autoimmunity induced. I shall explain why these three classes, regulated in the observed manner, provide a better system than a system consisting of just one class. CMI effector function is effective even when the target cell has a low density of recognized sites, whereas an effective attack by antibody (particularly IgG) requires recognition of a higher density of sites. It follows from this that CMI is more often damaging to a target cell. A cell with a low density of foreign sites is known to induce only CMI; this makes sense as antibody would be ineffective in attacking the cell and its presence might interfere with the efficient CMI attack. A cell with a high density of foreign sites induces only antibody; the antibody can successfully attack this target cell and so a CMI response is not required. The advantage of an absence of CMI is that any autoimmunity induced as a consequence of crossreaction with self antigens will be only antibody mediated, and antibody is less often damaging than CMI. The consequences of autoimmunity are thus minimized. Consider the problem of providing effective immunity and minimizing the consequences of autoimmunity if only one class were available. If this class was like antibody, it would be ineffective against cells with a low density of foreign sites, a most unsatisfactory situation as some virally infected cells could not be effectively attacked; if like CMI, all autoimmunity induced would be damaging to self as CMI is so effective against cells bearing only a low density of recognized sites.

These considerations suggest that the immune system has a mechanism for measuring the density of foreign sites on a cellular antigen; if low, a cell-mediated response is induced, if high, an antibody response. One possibility is that the number of helper T cells specific for the antigen provides this measure (14,15). This proposal leads to a clear prediction. An optimally immunogenic dose of an antigen with many foreign sites induces an antibody response in a normal animal as there are many helper T cells. In contrast, an antigen with few foreign sites induces a CMI response as there are few specific helper T cells. Consider administering an optimally immunogenic dose of an antigen with many foreign sites into an animal whose lymphocytes are systematically reduced in number. The number of helper T cells will be decreased and eventually a CMI response should be mounted in place of the humoral one. Consider an animal whose lymphocytes are sufficiently depleted that only a CMI response is mounted when it is challenged with an optimally immunogenic dose of antigen. The administration of additional specific helper T cells to such an animal should cause the coordinate suppression of the CMI and help the antibody response to the specific antigen, and leave responses to unrelated antigens unaffected. These predictions have been verified (16,17). They show that helper T cells switch a response from a cell-mediated to a humoral mode, illustrating the pivotal role these cells play in the process of determining the class of response an antigen induces. This principle appears to provide the beginning of a quantitative description of cellular interactions. I have purposely kept the analysis simple, but should state that the principle also accounts for the dependency of class induced on other variables, in particular the antigen dose (14,15).

(4) A role for accessory cells as integrators of environmental information?

A number of points arise if we accept the principle that the number of helper T cells [or more correctly helper T cell dependent signals (14)] determines the class of response an antigen induces.

(i) Cell interactions must be such that the response to epitopes e1,e2,....en, of a given entity, are roughly coordinately regulated, i.e. the same class is induced to all epitopes under given circumstances. For example, the response to a hapten depends crucially on the carrier. If we wish to induce CMI we couple the hapten to a carrier able to induce such a response, if we wish for IgM we couple it to a carbohydrate that induces this response, and if IgG to a very foreign carrier that provokes the production of IgG. These observations only apply if the coupling of the hapten is not so extensive as to change the nature of the carrier. Thus the class of response to the hapten is different in each of these cases and is determined by the nature of the carrier. The important property of the carrier in this respect appears to be its degree of foreigness.

(ii) An antigen specific for a precursor cell must deliver signals to this cell (either directly or indirectly) in a manner dependent upon its density of foreign sites. It is difficult to envisage how this can occur efficiently and reliably through a series of monospecific T cell-monospecific T cell interactions. This difficulty gives rise to a frankly speculative alternative: an accessory cell could passively absorb helper/inducer factors specific for foreign epitopes from a few or many lymphocytes, depending on the nature of the antigen, and deliver a signal to other cells, the intensity being dependent on the density of antigen-bound specific factors present on its surface. In this way the accessory cell could "integrate" information concerning the environment it is in and present this to appropriate cells.

(iii) The generalization that the density of foreign sites an antigen bears determines the class of response induced has mechanistic implications. It implies that the crucial signals determining the class induced occur through the linked recognition of antigen rather than through idiotypic-antiidiotypic interactions. This suggests, in the context of the proposal made in (ii) above, that antigen may be bound by both receptors on precursor cells and helper/inducer factors present on accessory cells. In this case one can envisage how the condition described in (i) above may be accomplished, as roughly the same signal will be delivered to precursors specific for e1,e2,....en.

(5) Antigen-receptor heterogeneity and the coherent regulation of the immune response: Can the rules of intercellular communication ensure that idiosyncratic lymphocytes act in unison?

A number of experimental observations suggest that the induction by antigen of lymphocytes, with receptors heterogenous for antigen, can be coherently and precisely regulated by external variables such as the concentration of antigen. This coherence, manifest as an attribute of the whole system, is surely physiologically significant, and must impose restrictions on possible models of cellular networks and intercellular communication.

One set of observations illustrating this coherence comes from Mitchison's estimate of the in vivo concentrations of the antigen, bovine serum albumin, that cause low-zone paralysis, namely 10^{-8} M, induction/sensitization for an antibody response, namely 10^{-7} M, and high-zone paralysis, occurring at 10^{-5} M (16). Thus only a thousand fold difference in concentration of antigen results in three different modes of sensitization, either negative or positive, at the humoral level. The binding constants of the receptors on the relevant B cells for this antigen probably differ amongst themselves by at least a thousand fold; the heterogeneity of B cell receptors would, in the context of simple models of cellular regulation, result in a broad dose-response curve. This poses the question of how regulation can achieve the coherent induction of a heterogenous population.

It is most probable that the coherence of B cell induction/inhibition, as a function of antigen concentration, reflects coherent induction/inhibition on the part of T cells, i.e. the preferential induction of suppressor T cells in low- and high-zone paralysis and helper T cells during sensitization. Coherence at the B cell level can thus be "explained" by removing the problem of "coherence despite heterogeneity" to the T cell level. The problem can be solved at the T cell level if there are two components in the network with distinct characteristics: (1) a requirement for a specified kind of interaction of antigen with specific cells or factors they produce before helper or suppressor T cells are induced, i.e. this interaction encodes the criterion for the kind of response required, and (ii) highly cooperative circuits that amplify the activity of the appropriate cells once a decision has been made to induce these cells. Highly cooperative systems allow a very sharp response to an external variable, as seen in the melting of ice as a function of temperature. One currently studied and probably relevant example in cellular immunology of a highly co-operative system is the autocrine growth of helper like T cells, e.g. the production of IL-2 by T cells that respond to this growth factor (17). Once high affinity cells respond they can help, in the presence of antigen, their lower affinity partners to respond as well. Such rules allow a heterogenous population of cells to act in unison.

I have described how some higher level properties of the immune system restrict potential descriptions of cellular interactions. I have not attempted to discuss how such considerations may be employed to suggest a coherent mechanistic scheme of cellular interactions. My attempts to achieve this are described elsewhere (14,15) and this was not the purpose of the exercise. Neither have I attempted to discuss alternative solutions to problems I have discussed. I should mention two such views as they are currently held by many and appear to be inconsistent with what I have suggested above. The strength of the interaction between an antigen (or its fragments) and class II molecules is postulated to frequently be the determining factor in whether a primary response is induced. Weak or nonimmunogenic antigens are either small in size or very similar to self. These properties are in no way related to their inability to interact with class II molecules. There are few helper T cells for such antigens, and it is reasonable to suppose that it is the sparcity of such cells that does not allow a response to be initiated, particularly in view of the requirement for specific T-specific T cell interaction in the induction of help (9). Secondly, it is sometimes suggested that a major contribution to the decision of whether helper or suppressor T cells are induced depends on the nature of the epitopes on the antigen, i.e. whether the epitopes "individually" belong to the helper or suppressor classes. I find this difficult to reconcile with the fact that even the same antigen, consisting of many epitopes, can induce either helper or suppressor T cells depending on the dose given, and that this effect of dose appears to hold for many antigens. In other words, I would argue that the help/suppression decision is not intrinsic to an epitope but governed by other principles as indicated above. Finally, the paradoxes surrounding IJ put the understanding we have in perspective. These mysteries at the molecular level hold promise that unrecognized attributes of the immune system still await discovery.

References

1. Hawkes, Terence. 1977. Structuralism and Semiotics. University of California Press, Berkeley and Los Angeles, CA.
2. Russell, B. History of Western Philosophy. G. Allen and Unwin, London, 1946.
3. Hume, D. 1739, 1740. Treatise of human nature.
4. Francis, T.T., Davenport, F.M. and Hennessy, A.V. 1953. Epidemiological recapitulation of human infection with different strains of influenza virus. Trans. Ass. Am. Physicians 66:107-117.
5. Asherson, G. and Stone, S.H. 1962. Selective and specific inhibition of 24 hour skin reactions in the guinea pig. I. Immune deviation:description of the phenomenon and the effect of splenectomy. Immunology 9:205-207.
6. Parish, C.R. 1972. The relationship between humoral and cell-mediated immunity. Trans. Rev. 13:35-66.
7. Challacombe, S.J. and Tomasi, T.B. 1980. Systemic tolerance and secretory immunity after oral immunization. J. Exp. Med. 152:1457-1471.
8. Bretscher, P.A. and Cohn, M. 1972. A theory of self-nonself discrimination. Science 169:1042-1049.
9. Bretscher, P.A. 1986. A cascade of T-T interactions, mediated by the linked recognition of antigen, in the induction of T cells able to help delayed-type hypersensitivity responses. J. Immunol. 137:3726-3733.
10. Nossal, G.J.V. and Pike, B. 1980. Functional clonal deletion in immunological tolerance to MHC antigens. Proc. Natl. Acad. Sc. USA 77:1602-1606.
11. Bretscher, P.A. 1977. An integration of B and T lymphocytes in immune activiation, in B and T cells in immune recognition, F. Loor and G. Roelants, Eds., John Wiley and Son.
12. Herzenberg, L.A., Tokuhisa, T. and Hayakawa, K. 1983. Epitope-specific regulation. Ann. Rev. Immunol. 1:609-632.
13. Cunningham, A.J. 1976. Self-tolerance maintained by active suppressor mechanisms. Trans. Rev. 31:23-35.
14. Bretscher, P.A. 1874. Hypothesis: On the control between cell-mediated, IgM and IgG immunity. Cell. Immunol. 13:171-195.
15. Bretscher, P.A. 1981. Significance and mechanisms of cellular regulation of the immune response. Fed. Proc. 40:1473-1478.
16. Mitchison, N.A. 1969. In regulation of the Antibody Response, B. Cinader, Ed., Thomas Springfield, Illinois, p.54.
17. Cantrell, D.A. and Smith, K.A. 1984. The interleukin-2 T-cell system: A new cell growth model. Science 224:1312-1316.

WHITHER THE IMMUNE RESPONSE? FACTORS DIRECTING THE RESPONSE TO HUMORAL OR CELL-MEDIATED IMMUNITY

Kim Bottomly
Howard Hughes Medical Institute
and Section of Immunobiology, Department of Pathology
Yale University School of Medicine
New Haven, CT 06510/USA.

The effector phase of an immune response is commonly described as being either humoral or cell-mediated. The pathological consequences of an immune response are largely determined by the nature of the effector pathway followed. However, the factors that determine which effector mode is activated are not understood. As this represents a major choice made by the immune system, and as the result of the choice has biological and theoretical consequences, studies in my laboratory have been addressing this issue.

In studying lymphocytes that participate in immune responses, it has been determined that helper T cells required for antibody responses to most antigens, and T cells mediating delayed-type hypersensitivity responses, are both $CD4^+$ and recognize protein antigen in the context of self class II major histocompatibility complex (MHC) antigens. For this reason, it appeared that the same T cell could perform both functions. However, the availability of a panel cloned T cell lines, all of which express CD4 and are antigen and class II restricted, has allowed a re-examination of this point. Three groups have recently reported that such $CD4^+$ T cells can be divided into two distinct types (1-4). One type produces interleukin 4 (IL4) as its autocrine growth factor, and is responsible for helping antigen-specific B cells to secrete antibody. Cells of this type do not mediate delayed type hypersensitivity reactions (Askenase and Janeway, personal communication; Cher and Mosmann, personal communication). Thus, I propose to refer to cells of this type as helper $CD4^+$ T cells. The other set secretes interleukin 2 (IL2) as its autocrine growth factor, and also secretes interferon gamma and lymphotoxin as characteristic lymphokines. Cells of this set, while able to participate in certain polyclonal B cell activation assays, do not provide help for antigen specific

NATO ASI Series, Vol. H23
The Semiotics of Cellular Communication in the Immune System
Edited by E. E. Sercarz et al.

primary B cells and are, in fact, suppressive to signals delivered by the $CD4^+$ helper T cell set. Cells of this set do mediate delayed type hypersensitivity responses, and also will kill if the target cell bears the appropriate class II molecules, presents the appropriate antigen, and is sensitive to the lytic mechanism of the T cells. I propose to refer to cells of this type as inflammatory $CD4^+$ T cells, Although not available in the mouse, the monoclonal antibody Ox22 has recently been shown Spickett and coworkers (3) to similarly subdivide normal rat $CD4^+$ T cells, although characterization of lymphokines produced by such cells is not yet reported.

Given that helper $CD4^+$ T cells are quite distinct from inflammatory $CD4^+$ T cells, and that the data in the rat suggest that this applies to normal populations as well, it seems likely that one answer to the question of regulation of the effector mode of an immune response lies in the requirements for activation and expansion of each set of cells during antigen priming in vivo. To examine this question, the growth requirements of cloned T cell lines typical of each set of cells have been examined. Both IL2 and IL4 can serve as T cell growth factors, as shown by the proliferative response of the cloned line HT-2 to these two lymphokines. Furthermore, it has been reported recently that IL-4 is the autocrine growth factor for some T cell lines, including those employed in these studies (5; J. Horowitz and C. Janeway, personal communication). Thus, in vitro proliferative responses of several cloned lines of cells representative of the two cell sets have been examined in the presence of purified IL2 or IL4. $CD4^+$ T cells of the helper type, which produce IL4, will proliferate in response to this lymphokine, while the cells of the inflammatory type will not. It should be noted that the proliferative response of $CD4^+$ T cells to IL4 is very dependent on the additional presence of the macrophage product IL-1; however, even in the presence of IL-1, proliferation of cells of the inflammatory set is not observed. When responses to IL-2 are examined, both sets of cells respond to recombinant derived IL-2 at essentially the same titer.

This difference in growth factor requirements between the two sets of cells suggests that priming regimens that induce activation of IL4 producing helper T cells will lead to selective expansion of this cell set, since only cells of this type can respond to IL4. The requirement for IL-1 suggests that macrophages are required for helper T cell priming as well, and this accords well with experimental data from many laboratories. Priming regimens that selectively activate T cells of the inflammatory type, and lead to IL-2 production, would be expected to expand T cells of both types, since both sets of cells respond to IL-2 by proliferating. However, inflammatory T cells produce interferon gamma as well as IL-2, and it should be noted that interferon gamma is a selective antagonist of IL-4 in many functional assays, so that expansion of IL-2 producing cells by IL-2 production, while it might also expand IL-4 producers, would also antagonize the functions of these cells, thus suppressing the humoral mode of the effector response. Thus, lymphokine balances deriving from two subsets of $CD4^+$ T cells might play a critical role in determining the modality of effector response that is elicited by an antigen.

What other characteristics of these two $CD4^+$ T cell types might be important in determining the mode of an immune response? This is now a critical issue to understand as it seems likely that the determination of effector mode will in turn rest on this distinction. What can one learn from studies on the selective activation of the two different $CD4^+$ T cell types? It is interesting to note that one situation in which priming of the two $CD4^+$ T cell types is altered is depletion of B cells in vivo prior to priming. Helper $CD4^+$ T cells in the spleen are primed normally in the absence of B cells, but priming for cell mediated immunity and for the T cell proliferative response is ablated. Interestingly, this effect of B cell depletion applies to the response to protein antigens, but not to the response to contactants, which is unaffected by B cell depletion. Thus, presentation to helper $CD4^+$ T cells may involve non-B cells, while presentation of protein antigens to inflammatory $CD4^+$ T cells appears to involve B cells. This may relate to the IL-1 dependence so easily observed with IL-4 responsive lines, and so difficult to observe with IL-2 producing lines. In addition, the inflammatory $CD4^+$ T cells may require high concentrations of antigen: class II MHC complexes for priming-levels achievable with soluble antigens only on antigen-specific B cells. Thus, co-factor and antigen:class II MHC density requirements for growth may also differ for the two types of $CD4^+$ T cells, perhaps reflecting differences in relevant antigen presenting cells. While it seems paradoxical that a T cell whose role is to activate B cells is primed by encountering antigen on a macrophage, yet a T cell whose role is to induce inflammatory responses is primed by encountering antigen on a B cell, these differences may reflect more the type of antigen studied than fundamental differences in antigen presentation. Most protein antigens studied by immunologists are given in bacterial adjuvants as soluble proteins, and used as soluble proteins for eliciting reactions. In fact, inflammatory T cells normally respond to infected cells, such as macrophages harboring viruses or intracellular bacteria, which may achieve very high levels of antigen. The actual control of cell-mediated as opposed to humoral effector modes is not clear, and remains an area of great interest. That one aspect may reflect the different growth factors used by the two cell types, may provide one link to understanding how such "decisions" are made in the course of an immune response. If one wishes to control that response, then all the missing links must be found. That the set of $CD4^+$ T cells that mediates one response is distinct from that which induces the other is a major step forward. A second step is discriminant growth requirements. The most important missing link now is finding how these sets are selectively enlisted in responses dominated by humoral or cell-mediated immunity.

References

1. Kim, J.A., A. Woods, E. Becker-Dunn and K. Bottomly. 1985. Distinct functional phenotypes of cloned Ia-restricted helper T cells. J. Exp. Med. 162:188.
2. Mosmann, T.R., H. Cherwinski, M.W. Bond, M.A. Giedin, and R.L. Coffman. 1986. Two types of murine helper T cell clone. I. Definition according to profiles of lymphokine activites and secreted proteins. J. Immunol 136:2340.
3. Spickett, G.P., M.R. Brandon, D.W. Mason, A.F. Williams and G.R. Woollett. 1983. MRC Ox22: a monoclonal antibody that labels a new subset of T lymphocytes and reacts with the high molecular weight form of the leukocyte-common antigen J. Exp. Med. 158:795.
4. Reinherz, E.L., C. Morimoto, K.A. Fitzerald, R.E. Hussey, J.F. Daley and S.F. Schlossman. 1982. Heterogeneity of human $T4^+$ inducer T cells defined by a monoclonal antibody that delinates two functional subpopulations. J. Immunol. 128:463.
5. Lichtman, A.H., E.A. Kurt-Jones, and A. Abbas. B cell stimulatory factor 1 and not interleukin 2 is the autocrine grown factor for some helper T lymphocytes. Proc. Nat. Acad. Sci. USA., In press.

IMMUNOSCHISMATICS: WHO IS GUILTY OF ORIGINAL ANTIGENIC SIN?

Charles A. Janeway, Jr.
Section of Immunobiology
Howard Hughes Medical Institute
at Yale University School of Medicine
New Haven, CT 06510

Original antigenic sin was a term coined by Francis (1) to describe the tendency of a person to produce anti-influenza virus antibody specific for the original immunizing virus upon re-immunization with a related but antigenically distinct virus. In this paper, I will examine this process in terms of the cleavage, by antigen, of an intact immune system into a responding and a suppressed compartment. The importance of this schismatic property of antigen acting upon an immune system in the interpretation of many phenomena, some of which might, I believe erroneously, be regarded as choices, will be discussed in some detail.

It is known to be a central characteristic of the immune system that prior exposure to antigen alters the subsequent response to that antigen; the enchanced response upon antigenic re-exposure is the property known as immunological memory, and reflects, at least in part, an antigen-specific expansion of both B and helper T cell compartments. That this state might be more complex was suggested by the finding of original antigenic sin. In this phenomenon, reimmunization with a related antigen led to an enhanced response to epitopes shared with the original antigen, but low or absent responses to normally immunogenic epitopes of the secondary challenging antigen (2). While this phenomenon could be viewed as the combined result of expansion of memory cells specific for shared antigenic epitopes, possibly accompanied by antibody-mediated suppression of the response to other epitopes on the same antigenic particle, this seems unlikely to be the whole story. Pierce and Klinman (3) have shown that the responses of primary B cells to haptens can be suppressed by irradiated T cells previously primed with hapten-carrier. This suppression was hapten-specific and genetically restricted by

NATO ASI Series, Vol. H23
The Semiotics of Cellular Communication in the Immune System
Edited by E. E. Sercarz et al.

Igh-linked genes. Interestingly, antigen-primed B lymphocytes were resistant to such suppression. A similar phenomenon, termed epitope-specific immune regulation, has been described by Herzenberg and co-workers (4). Finally, recent studies by Cancro (personal communication) using the original system of Francis and Fazekas de St. Groth, plus the experimental model of Pierce and Klinman, has shown that an Lyt-2^+ T lymphocyte can transfer an epitope-specific suppression that is Igh-restricted. Thus, although original antigenic sin affects B cell responses, the mediator appears to be a suppressor T cell. The target of this cell is not known.

Does original antigenic sin only affect B cell responses? The answer appears to be no. Experiments carried out by C.W. Pierce (5) and co-workers, and by Bennink and Doherty (6), have shown that exposure of an immune system to antigen both primed antigen-specific, MHC-restricted T helper or T cytotoxic effector cells and prevented responses of T cells to subsequent exposure to antigen in the context of non-self MHC. Where examined, this effect has been shown to be antigen-specific, and to be mediated by suppressor T cells (7). It is particularly important that exposure to antigen in the context of one MHC activates cells specific for that antigen:MHC complex, and apparently blocks subsequent activation of cells specific for the same antigen in the context of other MHC molecules. Thus, it would appear that such effects are mediated by anigen-specific suppressor cells, and that primed effector T cells are resistant to the effects of such suppressor cells. One might conclude from these experiments that effector T cells are subjected to "original antigen:MHC sin."

One can envision two models, with radically different consequences for the nature of T cell mediated antigen recognition, to account for the phenomenon of "original antigen:MHC sin." One could propose that T cell receptors are inherently specific for antigen:MHC complexes, and that suppression is specific for antigen alone, consistent with the finding that antigen binds to suppressor cells. In addition, one would have to propose that T cells can exist in two states of differentiation, suppressor-sensitive and suppressor-resistant, and that priming by antigen:MHC moves them from one state to the other. Alternatively, one could argue that the T cell receptor contains two functionally effective subsites, one for antigen and one for MHC, and that exposure to ligands occupying only one such site has effects other than activation, such as inducing a permanent suppressor-sensitive state.

As it has been very difficult to demonstrate any effect of antigen or of MHC alone on T cells, the latter model seems unlikely. For the present discussion, I will consider that T cells recognize only antigen:MHC complexes, and are subject to antigen-specific suppression. The next question one would then like to ask is, if original antigen:MHC sin exists, could "original MHC sin" exist also, and if so, what form might it take?

The situation that immediately comes to mind is the thymic effect. Stem cells of an F1 mouse, maturing in a parental thymus, respond to antigen only when presented in the context of that parent's MHC molecules, although the antigen presenting cells carry both parent's MHC molecules (8,9). This effect can be viewed as a selective expansion in the thymus of T cell precursors having specificity for antigen recognized in the context of self-MHC or, alternatively, as the loss of a part of the normal F1 T cell repertoire. A loss of repertoire would imply that T cells capable of suppressing responses to any antigen in the context of the MHC absent from the thymus, but present in the periphery, would exist in such chimeric mice. Indeed, it has been so argued by Smith and Miller, whose data support this notion (10). This effect, by analogy, could be termed "original MHC sin."

Are there other examples of original MHC sin, in which the mechanism might be more amenable to experimental manipulation? If one argues that original MHC sin is the result of F1 T cells passing through the parental thymus without encountering class II MHC antigens of the other parent until arriving in the periphery, and these autoreactive T cells being suppressed in the periphery by allele-specific suppressor cells, then one might argue that injection of parental T cells into an F1 recipient should activate allele-specific suppressor cells. In fact, this effect has also been described. Wilson and co-workers (11) inject parental, alloreactive T cells into F1 recipients and, within 24 hours, induce allele-specific suppressor T cells that block GVH reactions. The precise mechanism of this effect is not understood, but the fact that such suppressor cells, once activated, block lethal GVH induced by T cells of appropriate specificity of any strain of origin argues strongly that this suppression is allele, rather than receptor, specific. The experiments of Wilson and colleagues apply only to allo- reactions. For such a mechanism to apply to the thymic effect, it would have to act upon MHC-restricted T cells in an allele-specific fashion. Indeed, just such an effect has been claimed by Katz and co-workers (12). Perhaps even more impressively, Tada and co-workers (13) have injected autoreactive cloned T cells into F1 recipients, and observed suppression of antigen-specific resposes restricted to the class II MHC product recognized by the autoreactive cloned T cell line. This experimental system may accurately reflect the "thymic effect," in which F1 T cells lose the potential of recognizing antigen in the context of one parent's MHC molecule.

Immunoschismatics is the property of the immune system to cleave, in response to any reference antigen, into responsive and non-responsive components. The responsive component consists of memory B and MHC-restricted effector T lymphocytes and, most importantly for the present discussion, antigen-specific suppressor T cells able to inhibit responses of all but the primed effector cells. This property explains the phenomenon of original antigenic sin, and sets the task of identifying both the suppressor T cells and the property of suppressor resistance conferred upon B and effector T cells by the priming antigen. In addition, there appears to be a phenomenon of original MHC sin, whereby MHC antigens encountered in the thymus confer resistance to suppression whereas MHC antigens not encountered in the thymus, but prevalent in the periphery, induce suppression of all T cells recognizing antigen in the context of that MHC allele. If this construction is correct, the thymic effect is not the result of expansion of MHC-restricted precursors, but rather a result of the peripheral suppression of restricted, suppressor-susceptible pre-effector T cells. Thus, the property of the immune system, here called immunoschismatics, may apply not only to antigen but also to MHC, and may account for the findings in thymic chimeric mice. The identification and control of the cells responsible for the immunoschismatic effect could be very important both for understanding the biology of the immune system and for protection against disease. Finally, as immunoschismatics is a property of interacting cells in the immune system, the answer to the question posed in the title, namely, who is guilty of original antigenic sin, must be the system itself, rather than any individual component.

Acknowledgements

The author wishes to thank all those whose data he has abused in preparing this article, and Liza Cluggish for the patience to prepare it for publication.

References

1. Francis, T., Jr. Ann. Int. Med. 39:203, 1953.
2. Fazekas de St. Groth, S., and Webster, R.G. J. Exp. Med. 124:347,1966.
3 Pierce, S.K., and Klinman, N.R. J. Exp. Med. 146:509, 1977.
4. Herzenberg, L.A., Tokuhisa, T., and Hayakawa, K. Ann. Rev. Immunol. 1:609, 1983.
5. Pierce, C.W., Kapp, J.A., and Benacerraf, B. J. Exp. Med. 144:1382, 1976.
6. Bennink, J.R., and Doherty, P.C. J. Exp. Med. 148:128, 1978.
7. Pierce, C.W., and Kapp, J.A. J. Exp. Med. 148:1271, 1978.
8. Zinkernagel, R.M., and Doherty, P.C. Adv. Immunol. 27:52, 1979.
9. Schwartz, R.H. in Paul, W.E., ed. "Fundamental Immunology," Raven Press, N.Y., pp. 379-438, 1984.
10. Smith, F.J., and Miller, J.F.A.P. J. Exp. Med. 151:246, 1980.
11. Bellgrau, D., and Wilson, D.B. J. Exp. Med. 149:234, 1979.
12. Katz, D.H., Katz, L.R., and Bogowitz, C.A. J. Exp. Med. 153:407, 1981.
13. Sano, K., Fujisawa, I., Abe, R., Asano, Y., and Tada, T. J. Exp. Med. 165:1284, 1987.

EPITYPIC-IDIOTYPIC DOMINANCE AS AN EVOLUTIONARY ANSWER TO THE INTERCELLULAR COMMUNICATION PROBLEM

Eli E. Sercarz
Department of Microbiology
University of California, Los Angeles

Immunodominance is a ubiquitous occurrence in the immune system, the extensiveness of which has been insufficiently appreciated. Dominance is expressed at the level of recognition of antigenic and MHC determinants, at the level of competition between cells, and at the level of the organism. A partial listing of specific cases in which dominance plays a part is listed in Table 1. In the discussion to follow, I would like to consider dominance as an example of convergent evolution based upon the need for the intercellular regulatory machinery to be based on unambiguous signs rather than assuming it is the automatic consequence of natural selection, acting on diverse populations of lymphocytes.

The immune response depends on a carefully coordinated, multicellular collaboration, with the overriding element being specificity. Despite the secretion of several non-specific lymphokines by many of the cells involved, there is no wholesale effector cell activation: non-antigen specific reactions occur, but they are in the minority. (When specific partners in collaboration secrete soluble activators, they are generally directed towards their targets and have a minimal effective radius.) How is specificity maintained? It would be impossible if the typical lymphoid organ represented a "bag" of clonal lymphocytes each with its unique receptor fine structure and partner cells in collaboration had to find each other haphazardly. Although the cellular dynamics of each subpopulation isn't fully understood, there has been a growing awareness of the importance of orchestration of cellular position and function. An organismal Beethoven guarantees that the appropriate lymphoid partners will meet up for collaborative interaction.

The notion of survival of the fittest lymphocytes surely is a component of several developmental schemes of members of the system, e.g.. the mutation and affinity selection of developing B cells. A somewhat restrictive concept of "fitness" in this regard has been proposed (1), one applied to maturing B cell clones, that permits mutation without loss of function. It is an intriguing premise that fitness resides not in a state

NATO ASI Series, Vol. H23
The Semiotics of Cellular Communication in the Immune System
Edited by E.E. Sercarz et al.

Table 1.

Immune Hierarchies

(1) T suppressor cell responses are highly epitope-restricted.
(2) T helper/T proliferative responses display hierarchical dominance: usually cells recognizing one determinant will occupy a dominant position vis-a-vis others.
(3) Cytotoxic T cell responses also reveal clear Class I molecule restriction preferences.
(4) IdX predominant idiotypes are found on antibodies of different specificities.
(5) Such predominant idiotypes also appear on Ts.
(6) Ab responses to an Ag are often of a single isotype.
(7) A characteristic small set of clonotypes predominates in the antibody response to a multideterminant antigen.

---

of being but in an eventual promise of becoming. In fact, the triumph of the immune system is that each lymphocyte is as fit in a general sense as another in the absence of a stimulus. The test of "fitness" arrives as a confrontation with a particular antigenic or idiotypic determinant: if effective stimulation occurs only weakly, or if repeated mutations abort development of the clone, the retrospective judgment comes -- poor fitness. The system exemplifies the notion of collective fitness. Nevertheless, despite the virtue of the collective in thinking of preparedness for the response to an antigen, a much more autocratic immunodominance presides once antigen arrives on the scene.

The question arises why, among the near billion total lymphocytes (in a mouse) and the tens of thousands specific for a multideterminant antigen, that a very limited oligarchy of cells becomes dominant. Is there difficulty in allowing a thousand flowers to bloom? Thinking in collective terms, wouldn't a choice of hundreds or thousands of clones insure greater survivability?

From a socio-political perspective, highly pyramidal management of governmental entities has been the norm, although somewhat earlier in this century, commission forms of city government were much more popular than now. Lines of responsibility, accountability, clarity in decision-making favor the single leader. Alexis de Tocqueville observed that democracy and diplomacy make bad partners because the qualities needed for successful management of foreign affairs--secrecy, consistency, agility and the ability to make quick decisions--can more likely be carried out by an individual than by a group. Again, in the personal arena, from a plethora of possibly equally fit mates, one is chosen: preferential single pairing seems to best suit the needs and foibles of human survival. (The point can be raised, however, that it is very difficult in some domestic pairings to distinguish between faithful singularity and a type of hierarchical dominance where the nominal spouse is merely predominant).

In these examples, it is evident that non-dominance arrangements not only are feasible but have been explored. In the immune system however, the universality of dominance strongly suggests it as a crucial aspect permeating the affairs and interactions of every subpopulation. This can be taken to be a semiotic need for clarity and simplicity in the effective and unambiguous regulation of responses, in a situation where chaos would result from thousands of independent regulatory circuits. I will now consider four different cases of dominance and some of the possible causes for its establishment and its consequences.

(a) Dominance at the induction source: choices at the level of antigen-presenting cells. We have recently reviewed evidence in favor of the idea that dominance can be instituted at the antigen-presenting cell level (2). The most evident manifestation is the dominance hierarchy that is established among T cells following injection of the native form of a protein antigen. The expressed response is restricted to one or two dominant peptide specificities which the strain repeatedly seems to prefer, among many possibilities. Immunization with peptides representing primary sequences from the native molecule will therefore reveal a larger repertoire, which never sees the light of day after administering the native molecule.

Several cases presented in previous articles (2,3) all lead to the following model: the final peptide that will be presented to T cells in conjunction with class II will depend on the molecular context of the determinant, i.e., the form of the starting material. We conclude that as the antigenic molecule unfolds stepwise in the endosomic vesicle, the first site specific for class II molecular attachment (agretope) will be revealed. In the usual in vivo instance of limited antigen, there will be immediate binding by ambient class II molecules, and this initial encounter with Ia can be postulated to "guide the processing" of the entire molecule, possibly favoring the initial site by protecting it from enzymic attack, or by enhancing degradation at other sites. The concordance of a primary unfolding site on the antigen and the location of the initially revealed agretope should differ among MHC haplotypes and with various non-MHC "processing genes". The sum total of these effects will be that specific sites on the antigen molecule, possibly distant from other sites of T cell recognition, can play a decisive influence on the course of antigen processing and presentation, resulting in the observed dominance (4).

Since this type of hierarchical dominance at the level of the cell that presents antigen is so frequently reproducible within a strain, it cannot be treated as an accident. Rather it is likely to be one of the numerous devices that semiotically streamline the workings of the immune system. The narrower the repertoire that is activated at the source, the easier the subsequent regulatory problems.

(b) Epitope dominance and its relationship to idiotype selection. It is a surprising fact that fully half of the antibodies appearing in the antibody response to multi-determinant lysozyme (HEL) require the presence of TIP, the three amino terminal amino acids of HEL. In this situation, a heterogeneity of clones using different V_H and V_L genes is known to be initiated within this area, which by definition would all have the same 'fine specificity'. Many of these clones bear the predominant idiotype of the primary anti-HEL response, IdXE; concomitantly there also appear non-TIP dependent B cells (about 25% of which are $IdXE^+$), and we tentatively attribute this recruitment to idiotope-recognizing Th cells, whose engagement is a consequence of the $IdXE^+$ B cell activation. It is our hypothesis, stated earlier (5), that a semiotically sensible mechanism for enlisting IdXE-recognizing Th at the appropriate time would be to couple idiotypic recognition to the recognition of another entity. The usual thinking about corecognition has been limited to the class I and class II MHC molecules (6). However, other possibilities exist, such as lymphokine receptors, which appear at fixed times in cell maturation, or other differentiation markers such as CD-4 which have the valuable property of being largely restricted to functional subpopulations (also see Conclusions).

By whatever means, the coordinated appearance of $IdXE^+$ and/or TIP-dependent PFC comprising about 2/3 of all antibody-forming cells is the consequence of injection of a multideterminant antigen and was entirely unexpected. (Comparable studies have not been carried out with other proteins to discover whether this focused primary response is the rule or an exception. In the case of influenza hemagglutinin, a predominant idiotype (IdX) was again found (7). Surely the control by a single idiotypic system meets the requirements for coherence in regulation.) The anti-lysozymes bearing the IdX were directed against non-overlapping, unique determinants on lysozyme; this characterizes so-called "regulatory idiotopes" (8) whose existence can be postulated to be controlled by regulatory T cells (9). Presumably, B cells will only be promoted to the antibody producing stage by complementary IdX-recognizing T cells.

Further integration occurs through involving the suppressive T cell arm of regulation in these same idiotypic motifs. Most frequently, the same idiotopic IdX found within the predominant B cell idiotype, also characterizes the idiotypy of the Ts cascade (10,11). Recent evidence has implicated B cells as a source of these Ts idiotypic motifs (12), as might have been predicted. An intermediary, idiotype-recognizing T cell would have to be intercalated which could then "transmit" the idiotypic image by selecting a Ts bearing the internal image. This transmitting cell in many systems is the T suppressor inducer cell (see below, under c). There is a profound evolutionary wisdom in the establishment of this self-referential system in which the T and B cells regulate each other's activation and reactivities are based on complementarity of predominant idiotypic motifs. Through a mechanism such as this the system achieves a unified, coherent regulation that accomplishes one of the immunosemiotic requirements of reducing the complexity of regulatory possibilities.

(c) <u>Time and place: the example of suppression, and the escape from suppression by certain "dominant" T helper cells</u>

The position of intercellular encounter is a still poorly studied aspect of regulatory immunology. (The chances of meeting at Grand Central Station are considerably enhanced if both parties designate the shoe-shine stand at Track 29.) In suppressive systems, some evidence exists for a segregation in time and place. The functional localization of Tse to the spleen is also temporal, in that at a later time, Ts effectors can be detected in peripheral draining nodes. The regulatory effector cells exert a self-limiting function that itself is kept in stringent control. Thus, the inducer of Ts cells, the Tsi, which is required for stimulation of the effector Tse, is in turn the most sensitive target of the Tse (13). The design of the system seems to be that in the peripheral lymph nodes, although Tsi activity can be revealed, no Tse are initially evident. The

Tse, however, can be found in the spleen soon after i.v. or i.p. injection; they are only detectable in the peripheral draining lymph nodes at a later time following local injection. Why do the suppressor cells remain centrally located for so long and what stimulates their later trafficking to peripheral outposts? Apparently, in this way, during the early non-suppressive phase in the lymph nodes, an immune response can begin, unhampered by regulatory T cells. This may be a result of an assemblage of antigen-presenting cells in the spleen, but not the lymph nodes, that favors the induction of Ts effector cells; it could represent one indirect method of controlling the level of organismal activation by antigen.

The coexistence of dominant suppressor T cells and sensitive T helper cells may lead to no response at all to certain antigens. In fact, this has been described to be a major cause of Ir gene nonresponsiveness (14,15). However, all strains in antigenic models displaying suppression are not non-responders. In some cases certain Th have been shown to be non-susceptible to the suppressive system (16), suggesting that "dominant" Th may lie atop a hierarchy for regulatory reasons. This can be viewed in reverse (17), as a crucial factor in responsiveness to an antigen: if T cells directed against a dominant Th determinant are susceptible to suppression (usually the case for small antigens), there is "genetic" nonresponsiveness; if they are resistant to Ts, then a response ensues.

(d) Limitations in the chronically stimulated state: B cell clonal dominance

The response to a multideterminant antigen gets locked into a major expressed clonotype which can be produced for the lifetime of the animal (E. Benjamini, unpublished). Clonal dominance may more usually be pauciclonal dominance, where from one to three stable spectrotypes (on isoelectricfocusing) arising from large clones seem to predominate while contributions by minority clones to the serum antibody pool are not seen except when measuring total antibody. Upon injection of a secondary antigenic stimulus, and even after many subsequent injections, a dominant (set of) spectrotypes appears in the isoelectric focussing spectrum which is characteristic for that mouse. In fact, the specific band pattern of antibodies can provide an excellent means of identification of the animal (A. Miller and D.E. Kipp, unpublished). This clonal dominance has never been satisfactorily explained. In the case of a dominant anti-DNP clonotype, E9, studied extensively at the National Institute for Medical Research, Mill Hill, UK, in the early seventies, it wasn't of the highest affinity for DNP (18). A host of characteristics might contribute to the dominance, including binding affinity, doubling time of the cell, the density of receptor or Class II molecules on the cell surface, sensitivity to varous lymphokines or growth factors, resistance to suppression or escape from a regulatory network. Whatever the true cause(s), the organism's modus operandi apparently is to focus on a single or very limited set of representatives among the many possibilities offered. This is perhaps reminiscent of the young, naive newspaper reader amazed at how startling events are so well regulated that each day, that there can be one story fascinating enough to provide a headline! Just as the headline story may be sub-optimally newsworthy, or chosen by the editor for reasons unknown, the strategic necessity of dominance of particular clones remains an inexplicable but repeatable characteristic of the immune response.

We can try to arrive at an explanation for the evolution of this as a typical pattern. The key, I believe, stems from the observations that particular B cells tend to present certain T cell determinants much more favorably than others, in several systems (19,20,21). If we consider the specificity T on the antigen SEMIOTICS, it might be predicted that only B cells with a particular specificity say "C", will present "T" with high efficiency. This principle of "preferential partnership" between B and T cells of certain specificities (20) will also relate to the T cell dominance hierarchy. It may be that in the haplotype we are considering, "T"-specific T cell clones are the dominant specificity to arise. Therefore, the highly efficient C-T pairing will lead to a focus on anti-C antibody formation. Now, among the anti-C precursor B cells we can imagine some members with an affinity of 10^5, 10^6, 10^7 or 10^8 L mol^{-1}. According to commonly accepted principles (22), the persistence of a clone will be related to its affinity for antigen following mutation and a trait called "adaptability", meaning the ability of the cell to survive despite extensive mutations at various sites (1). One particularly adaptable trait that should enhance survival would involve the efficiency of antigen processing, perhaps one result of having high affinity receptors. It may be the case that certain clones process antigen better then others, placing "T" antigen in association with class II molecules on the cell surface in record time. If it were necessary to receive continued signals from the helper T cells to promote B cell maturation, the rapidly-processing, C-specific B cell that efficiently delivered determinant T to Ia molecules would be favored for this reason alone, in a pool of continually dwindling antigen.

To summarize the view suggested in accounting for clonal dominance, mechanisms of somatic mutation and affinity selection in the B cell compartment must be considered, as well as the preferential pairing rules for T-B collaboration: we presume that such situations, leading to very effective antigen presentation to T cells, and reciprocally optimal signalling to certain B cells, will conspire to result in clonal dominance.

General Conclusions: (a) Reducing the informational overload

The unmistakeable prevalence of certain motifs, an expression of dominance at so many points in the immune system, is one of its underlying themes. The transmission of unambiguous information must be a high priority requirement since the amount of potential information in the system is literally vast. Each set of interacting T and B cell subpopulations in the mouse possesses a potential repertoire of from 10^6-10^8 different specific receptors (or more).

In the face of so voluminous an information source, the immune system demonstrates its intelligence and discipline, by reducing the number of alternatives: chaos is avoided by evolving mechanisms to favor the hegemony of certain motifs. Its organization can be viewed as channeled by a semiotic S-code (23) ("a reductive network superimposed on the infinite array of events that could have taken place --- in order to isolate a few pertinent events"). Although various limitations in interactions are imposed on the cells, what is gained is the ability to transmit cogent, directed messages. As Eco points out, "the fewer the alternatives, the easier the communication" (23). Actually, there are a group of S-codes regulating intercourse between different members of the immune community. In addition to proscribing certain interactions and permitting others, these S-codes explicitly have a time (differentiation stage) and place framework. Thus, certain effector cells might require a stimulus of a late-acting factor produced by one type of T cell for further development. (For example, this may be the role of idiotype-recognizing T cells in their stimulation of B cells bearing predominant "regulatory" idiotypes.)

(b) The creation of signs for semiotic clarity. One effective means of organizing this information is to convert the external (antigenic) or internal (idiotypic) specificity elements into usable signs by coupling their recognition to that of stable surface molecules (e.g. MHC). The corecognition entity which must be recognized simultaneously, in this case the MHC molecule, serves as a "guidance molecule" that can specify the functional subset of lymphocytes that will be activated (6). As we have contended elsewhere (5), the MHC Class I and Class II molecules may not be the only entities which are utilized as semiotic corecognition structures in cell interaction: there are also the isotypic or allotypic markers on recognition receptors, as well as a large number of differentiation markers on the cell surface which may represent receptors for required growth factors or may perform an auxiliary role in recognition by the cell. When a particular growth factor receptor appears on the cell, it marks a specific differentiation stage that has been reached. It would be reasonable for the system to exploit these markers as corecognition elements, although this has not been demonstrated. Nevertheless, the function of differentiation markers such as CD4 and CD8 in corecognition has been increasingly studied.

The nature of the immune system's receptors, which contain information for recognition of diverse patterns, but also are themselves recognized by receptors on other lymphocytes, idiotypically, provides a rich context for cell-cell interaction. This is not a haphazard network of receptor-receptor interaction: the idiotypic complementarity can be visualized as creating individualized meaningful circuits that are brought into play by antigenic stimulation.

(c) Recognition, communication, activation: a teleonomic synthesis

Teleonomy in the sense of Monod (24) is one of the principle attributes of the living state, referring to the necessity of each characteristic quality emerging during evolution, almost as a purposeful development. From the teleonomic perspective, an efficient solution to the three problems of recognition, communication, and activation had to be sought for the immune system. The sign systems evolved to foster appropriate partnerships, providing the necessary informational content for this purpose. Signalling and cell activation, to insure both positive and negative regulation, require additional teleonomic devices, which may or may not be included within the same receptor structure. To a degree even greater than with other protein molecules, receptor heterodimers on lymphocytes possess a diversity of domains teleonomically designed for optimal performance and economy. The complexity of the assemblage on the T cell required for recognition and signalling is now appreciated and it is clear that a multiplicity of molecules subserve each function (e.g., see Eichmann article in this volume).

There are contrasting goals to the recognitive problem of establishing an extensive repertoire of potential responsiveness, and the communication problem of circumscribing cellular interactions to correct partner cells. In the first case, the recognition must be as broad as possible or at least as broad as necessary, while in the second, the requirements are for simplicity and non-ambiguity. The teleonomic challenge was immense: to incorporate both of these opposing goals along with a third signalling requirement which dictated that the same cell might need to be activated or suppressed at different times. Although all the devices which developed in response to this challenge are not fully described or understood, it was my purpose to highlight the necessity for clarity in communication and to suggest that the ubiquity of immunodominance mechanisms originated as a teleonomic response to this challenge.

It might be illustrative to consider an example of binary choice in a behavioral model. Although it is conceivable and presumably possible that simultaneous discourse on 2 topics could be followed along with ambidextrous note-taking, the idea seems ludicrous. It can be concluded that the human brain has not evolved to function on 2 parallel cognitive tracks at the same time, for good reason. In fact, far from being a limitation, a singular focus allows the conscious mind greater understanding, owing to the reverberations and harmonic interactions of the new information with previous knowledge and experience.

Another aspect of this dominance seems to prevail in the use of a language by a multilingual individual: when speaking German, the German cassette crowds out (dominates) over items in the French cassette, fortunately, making it possible to carry on a spirited conversation in German without innumerable French insertions. Functionally, immune communication leading to coherent responses to antigen thrives in a similar fashion when parallel tracks are avoided (response to too many epitopes/idiotopes).

I have reviewed only a few examples of dominance situations in lymphocyte interactions. Perhaps, as pointed out by Celada elsewhere in this volume, such studies of the organization of the immune system - in this case, dominance - can provide useful prototypes for other disciplines. In any event, it seems fascinating that the various aspects of dominance that characterize other arenas, became employed in the evolution of the delicately balanced immune system. It can be proposed that at several distinct points in the evolution of the system, semiotic necessities propelled the development of the current functional model of the immune system. This system manages to combine an enormous potential for diversity with the exigencies of real life, the necessity for concerted action upon antigenic demand, which in the affairs of cells and men, requires a restricted focus.

References

1. Manser, T., Wysocki, L.J., Gridley, T., Near, R.I., and Gefter, M.L. (1985). Immunol. Today 6,94-101.
2. Gammon, G., Shastri N., Cogswell, J., Wilbur S., Sadegh-Nasseri, S., Krzych, U., Miller, A., and Sercarz E. (1987). Immunol. Reviews, in press.
3. Cogswell, J.P., Gammon, G., Wilbur, S., Shastri, N., Miller A., and Sercarz, E. (1987). In Antigen Presenting Cells, (L.B. Schook, ed.) Alan R. Liss, Inc., New York, NY, in press
4. Sercarz, E., Wilbur, S., Sadegh-Nasseri, S., Miller, A., Manca, F., Gammon, G., and Shastri, N. (1986). In Progress in Immunology VI: Sixth International Congress of Immunology. 6, 227-237.
5. Sercarz, E.E. (1984). In Regulation of the Immune System. (H. Cantor, L. Chess, E.E. Sercarz, eds). Alan Liss, Inc. New York, NY, pp. 651-662.
6. Mitchison, N.A. (1980). In Strategies of Immune Regulation (E.E. Sercarz and A.J. Cunningham, eds.). Academic Press, New York, NY, pp. 121-133.
7. Clarke, S., Staudt, L., Gerhard, W., and Weigert, M. (1986). In 6th International Congress of Immunology. Abstracts, p. 95.
8. Bona, C.E., Heber-Katz, E., and Paul, W.E. (1981). J. Exp. Med. 153, 951-967.
9. Sercarz, E.E., and Metzger, D.W. (1980). Springer Seminars in Immunopathology. 3, 145-170.
10. Germain, R.N., Sy, M-S., Rock, K., Dietz, M.H., Greene, M.I., Nisonoff, A., Weinberger, J.Z., Ju, S-T., Dorf, M.E., and Benacerraf, B. (1981). In Immunoglobulin Idiotypes (C. Janeway, E.E. Sercarz and H. Wigzell, eds.). Academic Press, New York, NY. pp. 709-723.
11. Araneo, B.A., Yowell, R.L., and Sercarz, E.E. (1985). J. Immunol. 134, 1073.
12. Hayglass, K.T., Benacerraf, B., and Sy, M-S. (1986). Immunol. Today. 6, 179-183.
13. Green, D.R., Gershon, R.K., Eardley, D.D. (1981) Proc. Natl. Acad. Sci. USA. 78, 3819.
14. Araneo, B.A., Yowell, R.L., and E.E. Sercarz. (1979). J. Immunol. 123, 961-967.
15. Jensen, P.E., Pierce, C.W., Kapp, J.A. (1984). J. Exp. Med. 160, 1012.
16. Krzych, U., Fowler, A.V., and Sercarz, E.E. (1985). J. Exp. Med. 162, 311-323.
17. Sercarz, E.E. (1985). In Concepts in Immunopathology. Karger Medical and Scientific Publishers, Basel and New York. Vol. 3, pp. 61-73.
18. Askonas, B.A., and Williamson, A.R. (1972). Eur. J. Immunol. 2, 487.
19. Sercarz, E., Cecka, J.M., Kipp, D., and Miller, A. (1977). Annales d'Immunologie. 128C, 599-609.
20. Manca, F., Kunkl, A. Fenoglio, D., Fowler, A., Sercarz, E., and Celada, F. (1985). Eur. J. Immunol. 15, 345-350
21. Berzofsky, J.A. (1983). Surv. Immunol. Res. 3, 103.

22. Siskind, , G.W., and Benacerraf, B. (1969). In Advances in Immunology. (F.J. Dixon, Jr., and H.G. Kunkel, eds.). Academic Press, New York, New York. pp. 1-50.

23. Eco, U. (1979). A Theory of Semiotics, Indiana University Press, Bloomington, IN.

24. Monod, J. (1972). Chance and Necessity, Vintage Books, New York, NY.

SUBJECT INDEX

NATO ASI Series H

Vol. 1: **Biology and Molecular Biology of Plant-Pathogen Interactions.**
Edited by J.A. Bailey. 415 pages. 1986.

Vol. 2: **Glial-Neuronal Communication in Development and Regeneration.**
Edited by H.H. Althaus and W. Seifert. 865 pages. 1987.

Vol. 3: **Nicotinic Acetylcholine Receptor: Structure and Function.**
Edited by A. Maelicke. 489 pages. 1986.

Vol. 4: **Recognition in Microbe-Plant Symbiotic and Pathogenic Interactions.**
Edited by B. Lugtenberg. 449 pages. 1986.

Vol. 5: **Mesenchymal-Epithelial Interactions in Neural Development.**
Edited by J.R. Wolff, J. Sievers, and M. Berry. 428 pages. 1987.

Vol. 6: **Molecular Mechanisms of Desensitization to Signal Molecules.**
Edited by T.M. Konijn, P.J.M. Van Haastert, H. Van der Starre, H. Van der Wel, and M.D. Houslay. 336 pages. 1987.

Vol. 7: **Gangliosides and Modulation of Neuronal Functions.**
Edited by H. Rahmann. 647 pages. 1987.

Vol. 8: **Molecular and Cellular Aspects of Erythropoietin and Erythropoiesis.**
Edited by I.N. Rich. 460 pages. 1987.

Vol. 9: **Modification of Cell to Cell Signals During Normal and Pathological Aging.**
Edited by S. Govoni and F. Battaini. 297 pages. 1987.

Vol. 10: **Plant Hormone Receptors.** Edited by D. Klämbt. 319 pages. 1987.

Vol. 11: **Host-Parasite Cellular and Molecular Interactions in Protozoal Infections.**
Edited by K.-P. Chang and D. Snary. 425 pages. 1987.

Vol. 12: **The Cell Surface in Signal Transduction.**
Edited by E. Wagner, H. Greppin, and B. Millet. 243 pages. 1987.

Vol. 13: **Toxicology of Pesticides: Experimental, Clinical and Regulatory Perspectives.**
Edited by L.G. Costa, C.L. Galli, and S.D. Murphy. 320 pages. 1987.

Vol. 14: **Genetics of Translation. New Approaches.**
Edited by M.F. Tuite, M. Picard, and M. Bolotin-Fukuhara. 524 pages. 1988.

Vol. 15: **Photosensitisation. Molecular, Cellular and Medical Aspects.**
Edited by G. Moreno, R.H. Pottier, and T.G. Truscott. 521 pages. 1988.

Vol. 16: **Membrane Biogenesis.** Edited by J.A.F. Op den Kamp. 477 pages. 1988.

Vol. 17: **Cell to Cell Signals in Plant, Animal and Microbial Symbiosis.**
Edited by S. Scannerini, D. Smith, P. Bonfante-Fasolo, and V. Gianinazzi-Pearson. 414 pages. 1988.

Vol. 18: **Plant Cell Biotechnology.**
Edited by M.S.S. Pais, F. Mavituna, and J.M. Novais. 540 pages. 1988.

Vol. 19: **Modulation of Synaptic Transmission and Plasticity in Nervous Systems.**
Edited by G. Hertting and H.-C. Spatz. 457 pages. 1988.

Vol. 20: **Amino Acid Availability and Brain Function in Health and Disease.**
Edited by G. Huether. 487 pages. 1988.

NATO ASI Series H

MIX
Papier aus verantwortungsvollen Quellen
Paper from responsible sources
FSC® C105338

If you have any concerns about our products,
you can contact us on
ProductSafety@springernature.com

In case Publisher is established outside the EU,
the EU authorized representative is:
Springer Nature Customer Service Center GmbH
Europaplatz 3, 69115 Heidelberg, Germany

Printed by Libri Plureos GmbH
in Hamburg, Germany